LAND AND MARINE HYDROGEOLOGY

LAND AND MARINE HYDROGEOLOGY

Edited by

M. TANIGUCHI
Research Institute for Humanity and Nature (RIHN)
335 Takashima-cho, Kamigyo-ku, Kyoto 602-0878
Japan

K. WANG
Pacific Geoscience Centre, Geological Survey of Canada
9860 West Saanich Road, Sidney, B.C.
Canada V8L 4B2

T. GAMO
Division of Earth and Planetary Sciences, Graduate School of Science
Hokkaido University, N10 W8, Sapporo 060-0810
Japan

2003

ELSEVIER

Amsterdam – Boston – Heidelberg – London – New – York – Oxford
Paris – San Diego – San Francisco – Singapore – Sydney – Tokyo

ELSEVIER B.V.
Sara Burgerhartstraat 25
P.O. Box 211, 1000 AE Amsterdam, The Netherlands

First edition 2003
ISBN: 0-444-51479-1

Transferred to Digital Printing 2006

CONTENTS

PREFACE

This volume represents an effort to bring together communities of land-based hydrogeology and marine hydrogeology. The ten articles collected in this volume deal with hydrogeology in a broad range of tectonic and geological settings, from mid-ocean ridges to continental margins, from coastal groundwater channels to karst aquifers. They provide examples of how geophysical, geochemical, biological, and engineering approaches are used in various types of hydrogeological observations and modeling.

The chapters are in two groups. The first five chapters address the issue of submarine groundwater discharge or its opposite phenomenon of seawater invasion. This is where land hydrogeology and marine hydrogeology overlap. Submarine groundwater discharge is a rapidly developing research field. The Scientific Committee on Oceanic Research (SCOR) and the Land-Ocean Interactions in the Coastal Zone (LOICZ) Project of the International Geosphere-Biosphere Program (IGBP) have recently established a working group for this research. IASPO (The International Association for the Physical Sciences of the Ocean) and IAHS (International Association of Hydrological Sciences) under IUGG (International Union of geodesy and geophysics) also recently form a new joint committee "Seawater/Groundwater Interactions" to collaborate with oceanographers and hydrologists. Chapter 1 presents a comprehensive overview of the methodologies employed in assessing the amounts and rates of submarine groundwater discharge. Chapters 2 and 3 focus on different geochemical signatures of such discharge. Chapters 4 and 5 discuss its engineering and environmental implications. The other five articles introduce frontier research topics in more typical land and marine environments. We believe that the lessons and experience learned in these works are useful to all aspects of hydrogeology. Chapter 6, dealing with karst aquifers, demonstrates how special techniques can be developed to quantify groundwater flow in special environments. Chapters 7 and 8 discuss the biological aspects of fluids in sedimentary basins and submarine sedimentary formations, respectively. The last two chapters of the book review research work on vigorous fluid flow in subsea formations and their significance in global tectonics. Geochemical characteristics of hydrothermal activities at a number of active continental margins are reviewed in Chapter 9, and multidisciplinary geophysical constraints of the permeability of young igneous oceanic crust are summarized in Chapter 10. These two chapters deal with a variety of driving mechanisms for fluid flow in subsea formations: buoyancy driven free convection, tidally induced flow, flow induced by tectonic strain, flow due to sediment compaction, etc.

Most papers were initially presented at the international workshop on "Water in Deep Earth", which is a part of International Forum on "Water and Life". The International Forum was held from September 7 to 9, 2000 at Okayama International Center, Okayama, Japan. The forum was hosted by the Hayashibara Foundation. We wish to thank the organizers of the Forum, Prof. Emeritus Isamu Kayane, University of Tsukuba, Prof. Yugo Ono, Hokkaido University, Prof. Keitaro Yoshiwara, Japan Advanced Institute of Science and Technology, Prof. Emeritus Akiyoshi Wada, University of Tokyo, and Mr. Kazuya Masaki, Hayashibara Foundation.

The following colleagues critically reviewed various chapters of the volume: D. Deming, S. Ge, H. Kooi, O. Matsubayashi, W. S. Moore, M. Taniguchi, H. Villinger, T. Tokunaga, T. Gamo, T. Saito. Their time and efforts are deeply appreciated.

Finally, We would like to thank Drs. Femke Wallien and Ms Tonny Smit at Elsevier Publication for accepting the idea of this book.

Editors: Makoto Taniguchi
Kelin Wang
Toshitaka Gamo

Land and Marine Hydrogeology
Edited by M. Taniguchi, K. Wang and T. Gamo

Assessment methodologies for submarine groundwater discharge

M. Taniguchi[a], W. C. Burnett[b], J. E. Cable[c] and J. V. Turner[d]

[a]Department of Earth Sciences, Nara University of Education, Nara 630-8528 Japan; makoto@nara-edu.ac.jp

[b]Department of Oceanography, Florida State University, Tallahassee, Florida 32306 USA; wburnett@mailer.fsu.edu

[c]Department of Oceanography and Coastal Sciences, Louisiana State University, Baton Rouge, Louisiana 70803 USA; jcable@lsu.edu

[d]Centre for Groundwater Studies, CSIRO Division of Land and Water, Private Bag, PO Wembley, WA 6014 Perth, Australia; jeff.turner@per.clw.csiro.au

Submarine groundwater discharge (SGD) in the coastal zone is now recognized as a potentially significant material pathway from the land to the ocean. This article provides an overview on several methodologies used to estimate SGDs. Measurements of SGD using "manual seepage meters" show that consistent and reliable results can be obtained if one is aware of and careful to prevent known artifacts. New "automated seepage meters" will help to understand the hydrological and coastal oceanographic processes with longer-term and higher-resolution measurements. Direct measurements of SGD by seepage meters and piezometers in local areas may be scaled up to a regional basis by use of natural geochemical and geophysical tracers. Water balance estimates, although useful for rough estimates, are usually not very precise because the uncertainties in the various terms used to construct the balance are often on the same order as the groundwater discharge being evaluated. Estimates of SGD via analytical and numerical methods depend mainly on the evaluations of the thickness of the aquifers and representative hydraulic conductivities, of which well-constrained values are usually difficult to obtain except in very well-studied areas.

1. INTRODUCTION

Submarine groundwater discharge (SGD), which occurs as springs and seeps in near-shore areas, is one of the pathways of water discharge from land to the ocean in the global water cycle. Submarine springs are found in many parts of the world and some of these are large enough to provide fresh water for human needs *[Kohout*, 1966]. The slow seepage of groundwater that flows out along most shorelines of the world may be even more important volumetrically than discrete springs. While it is difficult to detect groundwater seepage through sediments, this diffuse input may occur over broad areas and deliver potentially significant amounts of fresh water and dissolved components to the world's coastal oceans.

The term "submarine groundwater discharge" has been used in different ways over the years [*Taniguchi et al.*, 2002]. The definition of SGD with or without recirculated seawater has been ambiguous in the literature [*Younger*, 1996]. In this paper, we use the term SGD to represent all direct discharge of subsurface fluids across the land-ocean interface. The exact definition of SGD is described in detail by *Taniguchi et al.* [2002].

The direct discharge of groundwater into the coastal zone has received increased attention in the last few years as it is now recognized that this process may represent a potentially important pathway for material transport. One of the outcomes of this recent interest in SGD has been the establishment of SCOR (Scientific Committee on Oceanic Research)/LOICZ (Land – Ocean Interactions in the Coastal Zone) Working Group 112 [*Burnett*, 1999; *Kontar and Zektser*, 1999]. They recognized the need to define further and improve the methodologies of SGD assessment. As a consequence, SGD assessment intercomparison exercises have been organized in several "flagship" coastal sites. These experiments are being performed in order to compare directly several independent methodologies. The aim is to develop standardized approaches for the evaluation of SGD in coastal zones.

Several methods for direct measurements of SGD have been employed, including seepage meters [*Carr and Winter*, 1980], piezometers, and geochemical/geophysical tracers. Indirect methods, including water balance methods, hydrograph-separation techniques, theoretical analyses and numerical simulations, are used widely for basin-scale estimations of groundwater discharge into the ocean.

SGD has been reviewed by *Fairbrige* [1966], *Stringfield and Legrand* [1969, 1971] and *Zektzer* [1973], however, no overview work has been done since 1970's. The objective of this paper is to review the most commonly used methodologies in estimating SGD: (1) seepage meters; (2) piezometers; (3) tracer methods; (4) water balance approaches; (5) hydrograph separation techniques; and (6) theoretical analysis and numerical simulations. Furthermore, we identify specific areas that require further analysis and study.

2. SEEPAGE METERS

Measurements of groundwater seepage rates into surface water bodies are often made using manual "seepage meters." This device was first developed by *Israelsen and Reeve* [1944] to measure water loss from irrigation canals. *Lee* [1977] designed a seepage meter consisting of an end section of a 55-gallon (208 liters) steel drum fitted with a sample port and a plastic collection bag (Figure 1). The drum forms an open0bottom chamber that is inserted into the sediment. Water seeping through the sediment will displace water trapped in the chamber forcing it up through the port into the plastic bag. The change in volume of water in the bag over a measured time interval provides the flux measurement.

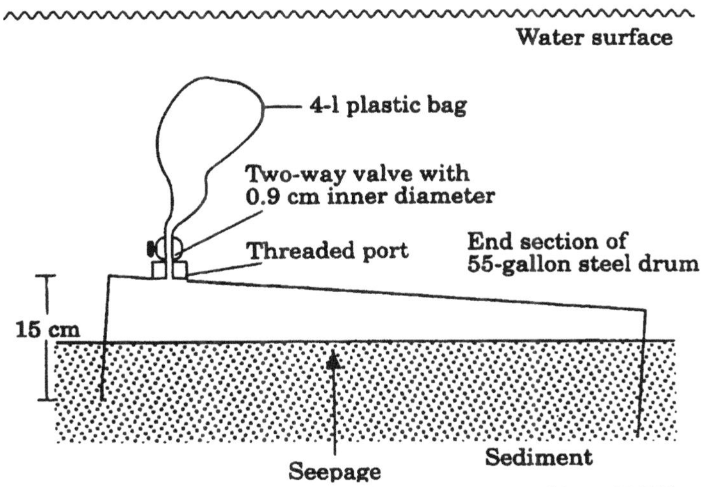

Figure 1. Lee-type manual seepage meter [*Lee*, 1977].

Studies involving seepage meters have reached the following general conclusions: (1) many seepage meters are needed because of the natural spatial and temporal variability of seepage rates [*Shaw and Prepas*, 1990a,b]; (2) the resistance of the tube [*Fellows and Brezonik*, 1980] and bag [*Shaw and Prepas*, 1989; *Belanger and Montgomery*, 1992] should be minimized to prevent artifacts; (3) use of a cover for the collection bag may reduce the effects of surface water movements due to waves, currents or streamflow activities [*Libelo and MacIntyre*, 1994]; and (4) caution should be applied when operating near the seepage meter detection limit [*Cable et al.*, 1997].

The influence on seepage measurements of frictional resistance in the tube connecting the drum to the collector bag produces inaccurate seepage rates. These inaccuracies can be minimized by using larger (0.9 cm ID) diameter plastic tubes [*Fellows and Brezonik*, 1980]. *Shaw and Prepas* [1989] found an anomalous short-term influx of water occurred when water was entering the collection bags after the bags were attached to seepage meters. They also showed that pre-filling the bags with a measured amount of water (about 1000 mL) reduced this measurement artifact substantially. The effect of surface water flow over the seepage meter and collection bag was discussed by *Libelo and MacIntyre* [1994], because it altered the hydraulic head within the meter augmenting seepage flow. Surface-water flow due to waves, currents, or streamflow reduces the hydraulic head in the meter resulting in an anomalously high value of measured seepage flux. Covering the collection bag with a bucket to isolate it from overlying surface water flow significantly reduces this error.

Although the application of manual type seepage meters has had some problems as mentioned above, a recent field evaluation of the "Lee-type" seepage meters showed that consistent and reliable results can be obtained if one accounts for these potential problems [*Cable et al.*, 1997]. Time-series experiments with empty collection bags demonstrated that the short-term anomalous influx reported by *Shaw and Prepas* [1989] was significant and could seriously affect the results, particularly at low seepage rates (Figure 2). Control experiments designed to measure the background seepage revealed a detection limit of 3 to 5 mL m^{-2} min^{-1} (0.4 to 0.7 cm day^{-1}) for their study. Despite these potential errors and detection limits, seepage meters provide useful measurements of groundwater seepage (including recirculated sea water [*Taniguchi et al.*, 2002]) in coastal waters, especially where the flux rates are high. They are commonly employed in lake settings to determine groundwater - surface water interactions (since the 1970's) and are increasingly utilized in coastal marine environments for studies where simple, inexpensive methods may be required.

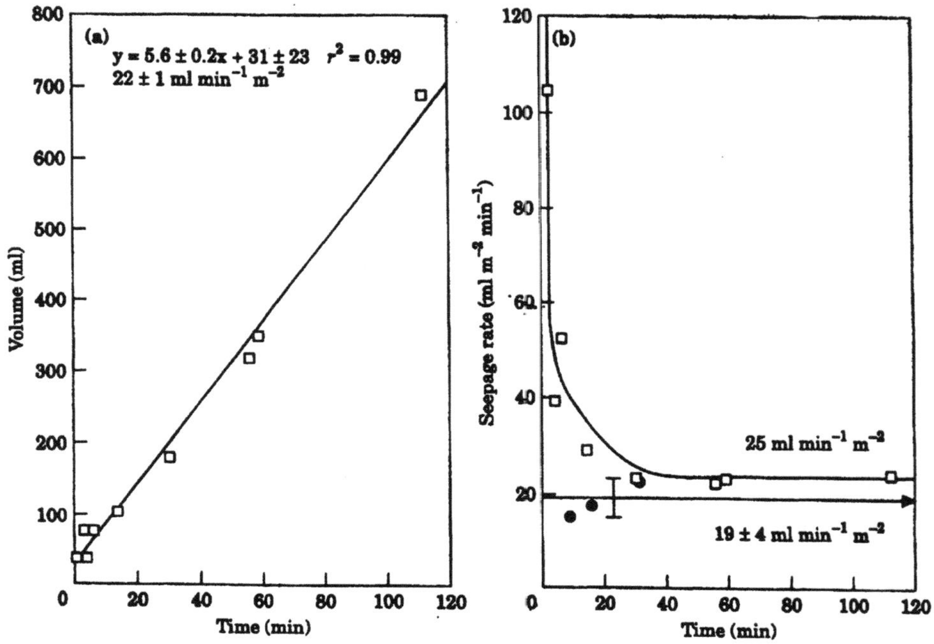

Figure 2. Observations of time-series measurements using seepage meters with empty (□) and 1000 mL pre-filled bags (●), plotted as : (a) volume (mL) vs. time (min.); and (b) seepage rate (mL min^{-1} m^{-2}) vs. time to demonstrate seepage meter response relative to collection time interval. Individual pre-filled bag measurements in (b) are shown together with the mean (arrow) [*Cable et al.*, 1997].

Perhaps the most serious disadvantage of using manual seepage meters in coastal zone studies is that they are very labor intensive. In order to obtain the groundwater discharge rate automatically and continuously, various types of automated seepage meters have been developed. Installations of seepage meters remotely from the surface of various water bodies are attempted by *Fukuo* [1986], *Cherkauer and McBride* [1988], and *Boyle* [1994]. *Sayles and Dickinson* [1990] constructed a seepage meter consists of a benthic chamber for the sampling and analysis of seepage through sediments associated with hydrothermal vents. Flow meters based on ultrasonic measurements are also used to evaluate seepage flow [*Paulsen et al.*, 1997]. Another example of such an automated approach for SGD measurements is the heat-pulse device described by *Taniguchi and Fukuo* [1993] and a similar meter constructed by *Krupa et al.* [1998].

The "Taniguchi-type (heat-pulse type)" automated seepage meter based on travel time of a heat pulse uses a string of thermistors in a column positioned above an inverted funnel covering a known area of sediment (Figure 3, *Taniguchi and Fukuo,* 1993). The basis of the method is measurement of the travel time of a heat pulse generated within the column by a nichrome wire induction heater. The travel time is a function of the advective velocity of the water flowing through the column. Thus, once the system is calibrated in the laboratory, measurements of seepage flow at a field site can be made automatically on a near-continuous basis. The Taniguchi meter has successfully measured seepage up to several days at a rate of about one measurement every five minutes [*Taniguchi and Fukuo*, 1996].

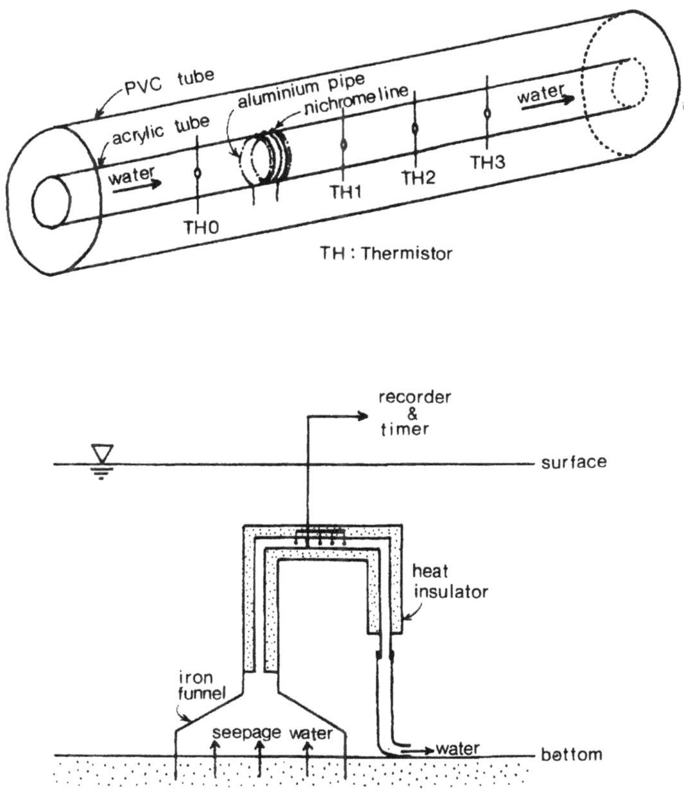

Figure 3. Taniguchi-type (heat pulse - type automated seepage meter [*Taniguchi and Fukuo*, 1993].

Taniguchi and Iwakawa [2001] recently developed "continuous-heat type automated seepage meter (Figure 4)". The mechanism of this meter is to measure the temperature gradient of the water flowing between the downstream (sensor A) and upstream (sensor B) positions in a flow tube with a diameter of 1.3 cm. The temperature gradient is caused by the heat continuously generated within the column. This method is so called "Granier method" (Granier, 1985). When there is no water flow, the temperature difference between sensors A and B in the column is the maximum, and it decreases with increasing the flow velocity. These automated devices should help us understand the temporal aspects of hydrological and coastal oceanographic processes with long-term and high-resolution measurements of SGD.

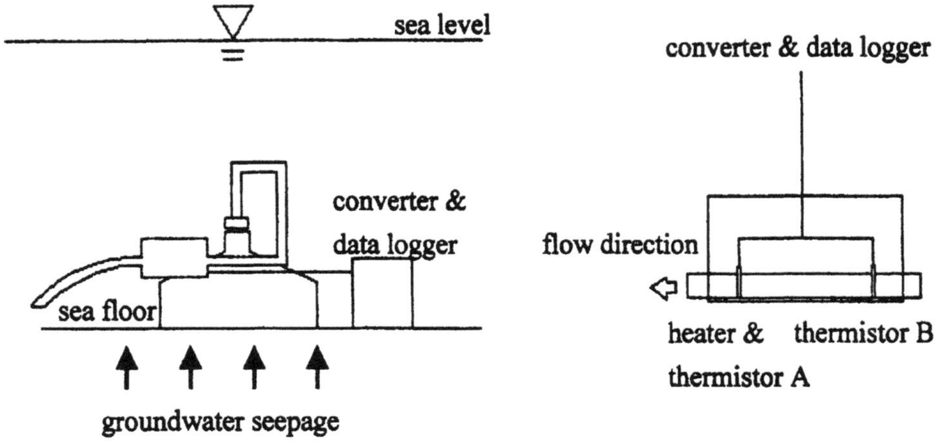

Figure 4. Continuous heat - type automated seepage meter [*Taniguchi and Iwakawa*, 2001].

3. PIEZOMETERS

Another method for assessing groundwater seepage rates is the use of multi-level piezometer nests. With this approach, the groundwater potential in the sediments can be measured at several depths [*Freeze and Cherry*, 1979]. Using observations or estimates of the aquifer hydraulic conductivity, one can then easily calculate the groundwater discharge rate into the ocean by using the one-dimensional form of Darcy's Law:

$$q = -K\,dh/dL \qquad (1)$$

where q is Darcian flux (groundwater discharge volume per unit area per unit time), K is hydraulic conductivity, h is hydraulic head and L is distance.

Similar to seepage meter measurements, piezometer nests suffer from the natural variability in seepage rates due to heterogeneity in the local geology. It is usually difficult to obtain representative values of hydraulic conductivity, which often varies over several orders of magnitude within an aquifer. Therefore, accurate evaluations of SGD using piezometers depend largely on the estimate of the aquifer hydraulic conductivity. Therefore, piezometer nests are often used in conjunction with seepage meters to estimate the hydraulic conductivity from observed seepage rates and the hydraulic gradient using equation (1) [*Barwell and Lee*, 1981; *Taniguchi*, 1995].

It is relatively easy to install piezometers in shallow-water sediments by direct pushing and hammering. Installing piezometers in deep-water sediments requires special techniques. A jetting procedure was used to install piezometers from a ship in relatively deep (15 m) water [*Cartwright et al.*, 1979]. However the jetting method leaves cavities in the sediments and introduces foreign water to the pore space around the piezometer screen, which could create inaccuracies in geochemical analysis of these pore waters [*Lee and Harvey*, 1996]. Another method for measuring the hydraulic gradient in deep water sediments is to position a probe that contains a differential pressure transducer into the sediment under the seabed [*Cartwright et al.*, 1979; *Schultheiss and McPhail*, 1986; *Davis et al.*, 1991]. However, this technique requires continuous monitoring of pore water pressure dissipation for hours to months, depending on sediment permeability, and the probe cannot usually penetrate beyond the upper 1 to 3 m of soft sediment. However, an underwater hammer operated from the surface can be used to install piezometers through 12 m of soft mud and 3 m of fine sand in a water column as deep as 30 m [*Lee and Harvey*, 1996]. In addition to the difficulties mentioned above, when piezometers are installed on the seafloor, lake bottom, or streambed, water may flow along the outer walls of the piezometer between the sediment and the piezometer, which would cause an underestimation of the actual SGD.

Although it is rare that more than one approach is employed to assess SGD in any one study [*Burnett et al.*, 2001], there are some exceptions [*Cherkauer and Taylor*, 1990; *Shaw et al.*, 1990; *Cherkauer and McKeregham*, 1991]. For example, *Shaw et al.* [1990] obtained groundwater seepage rates in a Canadian lake using four different methods: (1) a water budget; (2) wells (horizontal groundwater flow); (3) piezometers (vertical groundwater flow); and (4) seepage meters. The observed groundwater discharge rates using these different methods were consistent. However, this level of attention is certainly not always the case [*Woessner and Sullivan*, 1984].

4. TRACER METHOD

One approach for local to regional-scale estimation of groundwater inputs into the ocean is the use of naturally-occurring geochemical tracers. An advantage of groundwater tracers is that they present an integrated signal as they enter the marine water column via pathways in the aquifer. Small-scale variability is a serious problem for the use of seepage meters or piezometers, but such small-spatial-scale variations tend to be smoothed out over time and space in the case of tracer methods [*Burnett et al.*, 2000]. On the other hand, natural tracers require that all other tracer sources and sinks except groundwater be evaluated, an often difficult exercise. In order to standardize assessment methodologies, intercomparisons between various approaches, including seepage meters and natural tracers, are needed to evaluate the variability and deviation of SGD assessments.

Natural geochemical tracers have been applied in two ways to evaluate groundwater discharge rates into the ocean. First is the use of geochemical tracers enriched in the groundwater relative to the seawater. In other words, the concentration of a solute in the receiving water body is attributed to inputs of that component derived only from groundwater [*Moore*, 1996; *Cable et al.*, 1996a,b]. A second approach is the use of vertical profiles of the geochemical compositions in sediment pore waters under the assumption that its distribution can be described by a vertical one-dimensional advection-diffusion model [e.g. *Cornett et al.*, 1989; *Vanek*, 1993]. However, this is usually limited to the case of homogeneous media.

Over the past few years, several studies have employed the use of natural uranium decay-series nuclides ^{226}Ra and ^{222}Rn to assess groundwater discharge into the ocean [*Burnett et al.*, 1990, 1996; *Ellins et al.*, 1990; *Moore*, 1996; *Rama and Moore*, 1996; *Cable et al.*, 1996a,b; *Moore and Shaw*, 1998; *Corbett et al.*, 1999; *Hussain et al.*, 1999]. Ideally, in order to provide a detectable concentration, natural geochemical tracers should be greatly enriched in the discharging groundwater relative to coastal marine waters, conservative, and easy to measure. While radium and radon meet these criteria fairly well, other tracers exist which may be exploited for groundwater discharge studies. Some particularly attractive tracers are tritium and helium isotopes [*Burnett, et al.,* 2001] and major ions and strontium isotopes [*Martin and Gordon*, 2000; *Martin et al.*, 2000]. In applying geochemical tracer techniques, several system parameters must be assessed or defined, including geometric conditions (i.e., area, volume), water and constituent sources and sinks, residence times of the surface water body, and concentrations of the tracer. Sources may include ocean water, river water, groundwater, precipitation, *in situ* production, lateral water transport, sediment resuspension, or sediment diffusion. Sinks may include *in situ* decay or consumption, horizontal water column transport, horizontal or vertical eddy diffusivity, and atmospheric evasion. Through simple mass balances or box models incorporating both sediment advection and water column transport, the geochemical approach can be quite useful in assessing submarine groundwater discharge [*Cable et al.*, 1996b].

Methane (CH_4) is another useful geochemical tracer that can be used to detect SGD. Both ^{222}Rn and CH_4 were measured along the Juan de Fuca Ridge as a mean of estimating heat and chemical fluxes from the hydrothermal vents of that area [*Rosenberg et al.*, 1988]. Both ^{222}Rn and CH_4 were used to evaluate SGD in a coastal area of the northeastern Gulf of Mexico [*Cable et al.* ,1996a]. ^{222}Rn and CH_4 inventories in the water column and seepage rates measured using a transect of seepage meters were evaluated over several months within a shallow water location. The linear relationships between tracer concentrations and measured seepage fluxes were statistically significant (Figure 5). The inventories of ^{222}Rn and CH_4 in the coastal waters varied directly with groundwater seepage rates and had a positive relationship (95 % C.L.). In addition, water samples collected near a submarine spring in the same area displayed radon and methane concentrations inversely related to salinity and considerably greater than those found in surrounding waters. In a corresponding study, *Bugna et al.* [1996] demonstrated that groundwater discharge was an important source for CH_4 budgets on the inner continental shelf of the same region. In another example, *Tsunogai et al.* [1999] found methane-rich plumes in the Suruga Trough (Japan) and postulated that the plumes were supplied from continuous cold fluid seepage in that area.

Interactions between surface waters and groundwaters have been evaluated for many years using stable isotope tracers, such as ^{18}O and deuterium [*International Atomic Energy Agency*, 1979; *Krabbenhoft and Anderson*, 1986; *Turner and MacPherson*, 1990; *Krabbenhoft et al.*, 1990a,b; *Shimada et al.*, 1993; *Katz et al.*, 1995; *Katz et al.*, 1997; *Turner and Barnes*, 1998] and natural solutes such as [*Stauffer*, 1985; *Shedlock et al.*, 1993; *Vanek*, 1993; *Katz et al.*, 1997; *Martin and Gordon*, 2000; *Martin et al.*, 2000]. Their usefulness is attributed to the difference in isotope ratios or solute concentrations between sources (groundwater and surface water) due to the difference in their origin and interactions caused by hydrological processes such as evaporation. As with any other geochemical tracer, natural solutes are only effective if they are conservative or their sources and sinks can be quantified. For example, chloride makes an effective tracer for groundwater because it is conservative, and because groundwater with very low Cl^-, reduces seawater Cl^- concentrations

[*Martin et al.*, 2000]. In order to estimate regional scale SGD using tracers, the electrical conductivity of bottom sediments was mapped as a reconnaissance technique that allows effective evaluations of SGD with additional measurements by piezometers or seepage meters [*Harvey et al.*, 1997].

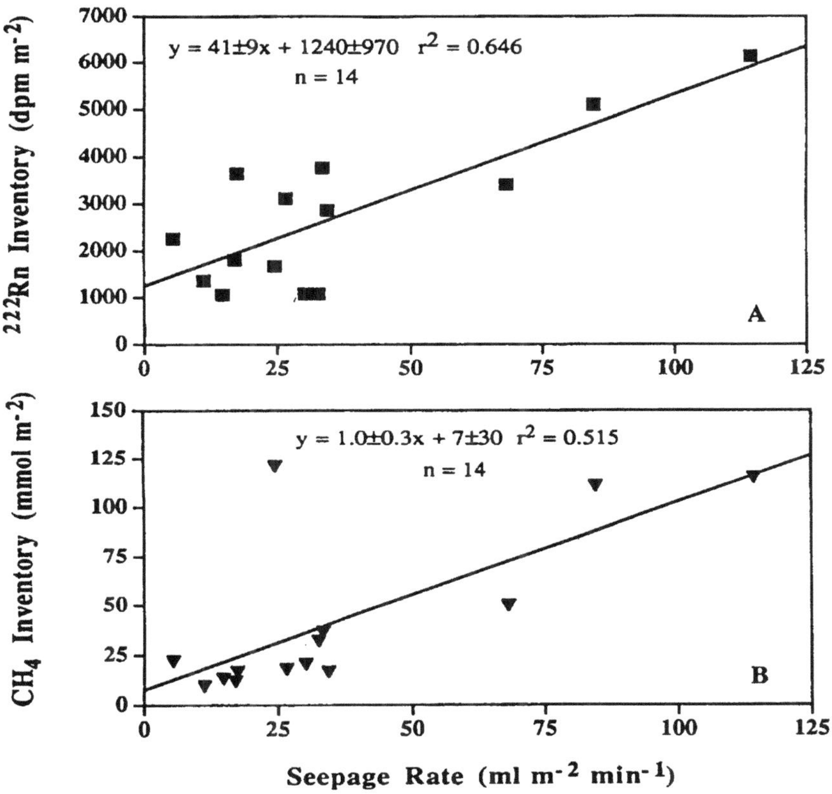

Figure 5. Relationship between ^{222}Rn and CH_4 inventories in the overlying water column and groundwater fluxes measured at one station by seepage meters in the coastal Gulf of Mexico [*Cable et al.*, 1996a].

In addition to geochemical tracers, geophysical tracers such as groundwater temperature can be used to estimate groundwater discharge rates. Two basic methods are employed when using temperature as a tracer: (1) temperature-depth profiles under the assumption of conservative heat conduction-advection transport; and (2) use of temperature differences in the groundwater - surface water system as a qualitative signal of groundwater seepage using techniques such as infrared sensors or other remote sensing methods.

Temperature-depth profiles in boreholes have been widely used to estimate groundwater fluxes because heat in the subsurface is transported not only by conduction but also by advection due to groundwater flow. The type-curve method was developed by *Bredhoeft and Papadopulos* [1965] for estimating one-dimensional groundwater fluxes based on a steady state heat conduction-advection equation derived from *Stallman* [1963]. This method has been widely used to estimate one-dimensional vertical groundwater fluxes [e.g. *Cartwright*, 1979; *Boyle and Saleem*, 1979; *Jessop and Vigrass*, 1989], one-dimensional horizontal groundwater fluxes [e.g. *Cartwright*, 1970; *Sakura*, 1977], and one-dimensional vertical groundwater fluxes with the effect of horizontal groundwater fluxes [*Lu and Ge*, 1996]. Simultaneous movement of one-dimensional transient heat and steady water flow was analyzed observationally [*Sillman and Booth*, 1993; *Constantz et al.*, 1994], numerically [*Lapham*, 1989], and theoretically [*Suzuki*, 1960; *Stallman*, 1965; *Taniguchi*, 1993; 1994]. The relationship between two-dimensional subsurface temperature and groundwater flux was theoretically analyzed by *Domenico and Palciauskas* [1973], and *Smith and Chapman* [1983]. More recently, surface warming caused by global warming and urbanization [*Taniguchi et al.*, 1999a] or deforestation [*Taniguchi et al.*, 1999b] was used as a signal to estimate groundwater fluxes (Figure 6). Borehole temperature data near the coast was also used for estimations of SGD into Tokyo Bay, Japan [*Taniguchi et al.*, 1998] and a saltwater-freshwater interface in Toyama Bay, Japan [*Taniguchi*, 2000]. All of these studies suggest that temperature-depth profiles in the coastal zone can be used to evaluate SGD.

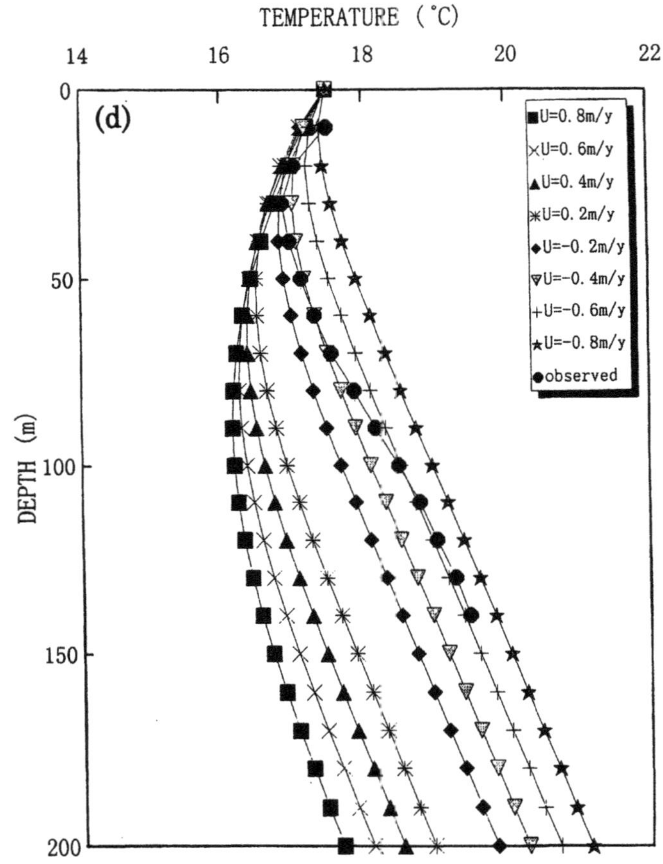

Figure 6. Observed and calculated temperature-depth profiles using heat conduction-convection equation to estimate upward groundwater fluxes (groundwater discharge rates) near Tokyo Bay. Negative U shows the upward groundwater flux [*Taniguchi et al.*, 1999a].

In order to evaluate SGD for regional-scale by using temperature as a signal, an infrared sensor was used in many areas [*Fischer et al.*, 1964; *Banks et al.*, 1996; *Roxburgh*, 1985; *Bogle and Loy*, 1995]. However, SGD values were not evaluated quantitatively though the locations of SGD were evaluated. These detectable locations are attributed to the spatial and temporal variation of both seawater and groundwater temperatures, which requires intensive field calibration. The use of remote sensing technologies to identify and quantify SGD is clearly an area for future research.

In scaling-up SGD measurements from point or local scales to the regional scale, it is important to evaluate the relationship between representative SGD data and the observed water depth. Results from deeper waters are not nearly as common as shallow waters but may be more representative of the mean result over wide areas. Methods used for scaling-up from local to regional as applied in the field of meteorology may be applicable to SGD, although the inherent heterogeneity of the aquifers must be addressed.

5. WATER BALANCE APPROACHES

The water balance equation for a basin has been used to estimate SGD and is described as follows:

$$P = E_T + D_S + D_G + dS \qquad (2)$$

where P is precipitation, E_T is evapotranspiration, D_S is surface runoff, D_G is groundwater discharge, and dS is the change in water storage. Over extended periods (i.e., years), dS is usually assumed to be negligible. Therefore, one needs to know precisely the precipitation, evapotranspiration and surface runoff for an accurate estimation of D_G.

Basin-scale estimations of SGD by a water balance method have been performed in many places, e.g., Perth, Australia [1.0×10^8 m^3/y; *Allen* 1976], Santa Barbara [1.2×10^5 m^3/y; *Muir* 1968], Long Island, New York [2.5×10^7 m^3/y; *Pluhowski and Kantrowitz*, 1964], and in the Adriatic Sea [1.7×10^{11} m^3/y; *Sekulic and Vertacnik*, 1996]. When both the area and volume of SGD are known, one can calculate the SGD flux. For example in the case of the Adriatic Sea [*Sekulic and Vertacnik*, 1996], the mean SGD flux of 0.68 m/y is calculated from the estimated SGD volume and the discharge area. Often, the area over which SGD occurs is unknown. Therefore, SGD volume or sometimes "volume of SGD per unit length of shoreline" [*Robinson*, 1996; *Sellinger*, 1995] is used for water balance studies, making it difficult to compare with the observed (local) SGD estimates shown as Darcy's flux (e.g., cm^3/cm^2/s, cm/s, m/y).

Water budget calculations, while relatively simple, are often imprecise for groundwater discharge estimations because uncertainties associated with values used in the calculations are often of the same order of magnitude as the discharge being evaluated. For instance in the global water budget constructed by *Garrels and MacKenzie* [1971], the estimated SGD is about 6 percent of estimated evaporation from the land, which is about the same as the uncertainty in evaporation estimates [See also *Burnett et al.*, 2001].

From a study designed to test the effects of climate change on groundwater discharge, *Oberdorfer* [1996] concluded that use of a water budget is an adequate first step in for assessing expected changes in simple groundwater basins. On the other hand, numerical modeling provides a more quantitative estimate of climate change perturbations when dealing with basins characterized by multiple sources and sinks.

In the case of basin-scale SGD estimation using water balance methods, SGD is the same as net groundwater discharge, D_G [*Zektser, et al.*, 1983]. On the other hand, SGD estimations based on smaller-scale studies, using tracers or seepage meters, include both net (fresh) groundwater discharge, and recirculated seawater [Taniguchi *et al.*, 2002]. Therefore, one should ideally separate SGD into two components in the cases of regional or local scale SGD estimation, because some sub-cycles of the water near the coastal zone exist in the basin-scale water balance. Another water balance approach using a budget based on the change in soil moisture has been performed for Tomales Bay, California [*Oberdorfer et al.*, 1990]. Their result was comparable to the result obtained by more traditional water balance estimations.

6. HYDROGRAPH SEPARATION TECHNIQUES

The hydrograph separation technique is based on the assumption that the amount of groundwater entering streams can be obtained by separating the potion from baseflow. This estimate may be extrapolated to the coastal zone. This technique was used by *Zektser and Dzhamalov* [1981] for the Pacific Ocean rim, by *Boldorski* [1996] in eastern Russia, by *Williams and Pinder* [1990] in the local coastal plain stream in South Carolina, and by *Zektser et al.* [1973] for global-scale estimation of SGD. Two approaches were used to separate the hydrograph for estimating the groundwater flow component. The first method is simply to assign a baseflow according to the shape of the hydrograph. This technique can be performed in several ways including the unit graph method [*Bouwer*, 1978; *Zektser et al.*, 1973]. However, a problem with this simple approach is evaluating baseline conditions that often change with time, space, and prevailing hydrological conditions. The hydrograph separation technique for world-wide SGD estimates [*Zektser et al.*, 1973] applies only to coastal areas with well-developed stream networks and to zones of relatively shallow, mainly freshwater aquifers.

The second method of hydrograph separation is the use of geochemical end-member concentrations. Usually, water and geochemical mass balances in a river are shown as follows:

$$D_T = D_S + D_G \tag{3}$$

$$C_T D_T = C_S D_S + C_G D_G \tag{4}$$

where D and C are the discharge rate and geochemical concentrations, respectively, and subscripts $_T$, $_S$ and $_G$ represent the total, surface water and groundwater components. From those two equationswith measured D_T, C_T, C_S, and C_G, we can solve for the two unknown values, D_S and D_G.

Recently, not only surface water – groundwater separation [*Fritz et al.*, 1976], but also the separation of three water components, namely groundwater, surface water and soil water, has been studied by using three different compositions with these end-members [*Tanaka and Ono*, 1998]. This method may also be applicable for separation of SGD into the fresh, mixing, and seawater components of SGD if one can identify tracers with sufficient sensitivity and resolution.

As with the water balance method, the uncertainties in the hydrograph separation terms are often on the same order of magnitude as the discharge being evaluated. For instance, the estimation of groundwater discharge in central and eastern European countries showed the average of estimated groundwater discharge (6 % of total water flow) is about 12 percent of the estimated evaporation [*Zektser and Loaiciga,* 1993]. This estimate is about the same order of uncertainty usually assigned to evaporation estimates.

Another problem of the hydrograph separation for estimating direct groundwater discharge into the ocean is that gauging stations for measuring the discharge rate in rivers are always located some finite distance upstream from the coast. Therefore, the groundwater discharge downstream of the gauging station is excluded [*Buddemeier*, 1996].

7. THEORETICAL ANALYSIS AND NUMERICAL SIMULATIONS

Offshore seepage rates from surficial unconfined aquifers have been described by an exponentially decreasing function with the distance from the coast [*McBride and Pfannkuch*, 1975]. They investigated the distribution of groundwater seepage rate through lake beds using numerical models. *Bokuniewicz* [1992] questioned the use of an exponentially decreasing function, and developed an analytical solution for SGD from steady state Richards' equation as follows;

$$q = (Ki/\pi k)\ \ln[\coth(\pi xk/4l)] \qquad (5)$$

hydraulic gradient, k is the square root of the ratio of the vertical to the horizontal hydraulic conductivity, l is aquifer thickness and x is the horizontal distance from the shoreline. *Bokuniewicz* [1992] concluded that an exponentially decreasing function would underestimate the SGD both near-shore and far from shore, and overestimate the SGD at intermediate distances. This difference between the exponential approximation and the analytical solution is similar to the contrast between an exponential representation and the numerical examples calculated by *McBride and Pfannkuch* [1975].

Theoretical analysis of groundwater seepage rates also has been made by *Fukuo and Kaihotsu* [1988] using conformal mapping techniques for areas with a gentle slope into surface water bodies (Figure 7). They found that most of the groundwater flows through a nearshore interface between surface water and groundwater. Equipotential and stream lines in the near-shore vicinity of the aquifer and the distribution of specific discharge through the sediment with different slopes demonstrate this point [Figure 7a; *Fukuo and Kaihotsu*, 1988]. Analytical solutions indicate that SGD decreases exponentially with distance from the coast and that the rate of decrease is greater when a gentler slope is present (Figure 7b).

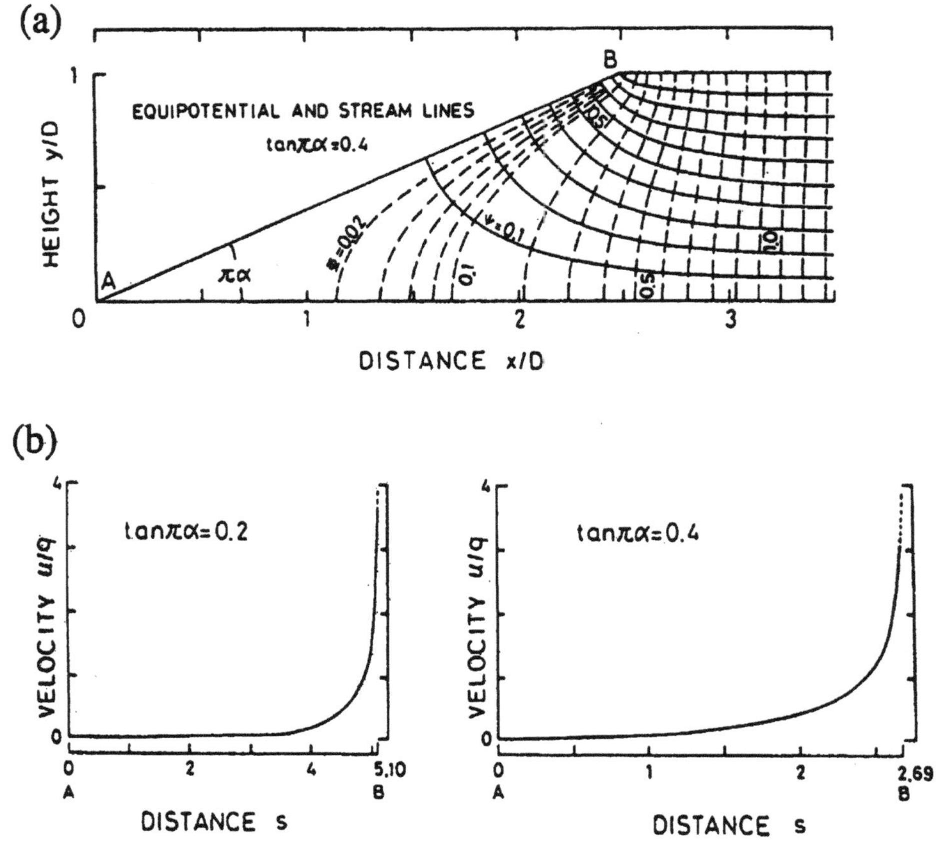

Figure 7. (a) Equipotential and stream lines in the vicinity of the sediment; and (b) distribution of specific discharge on the sediment surface with a gentle slope. The x and y axis in (b) are non-dimensional distance and velocity [*Fukuo and Kaihotsu*, 1988].

Interactions between surface waters and groundwaters also have been studied numerically by *Winter* [1976, 1983, 1986], *Anderson and Chen* [1993] and *Nield et al.* [1994]. *Linderfelt and Turner* [2001] numerically evaluated the net groundwater discharge to a saline estuary. *Smith and Turner* [2001] numerically evaluated the role of density-driven re-circulation component in the overall groundwater discharge to the same saline estuary.

Although deterministic modeling using packages such as MODFLOW [*McDonald and Harbaugh*, 1984] is common in the analysis of basin-scale groundwater hydrology, it has certain limitations. For example, spatial and temporal variations in boundary conditions are required for hydrological modeling, but this information is often hampered by our inability to acquire adequate field data within the time frame of a typical study. In addition to this, aquifer systems are usually heterogeneous, and it is difficult to assign representative values to hydraulic conductivity and porosity to characterize this heterogeneity. It is especially difficult to obtain representative values for hydraulic conductivity, which often vary over several orders of magnitude within short distances. However, recent stochastic or probabilistic model for groundwater studies can characterize the large spatial variability of parameters.

Modeling approaches for assessment of SGD include analytical or numerical hydrogeologic computations of groundwater discharge, combined hydrological and hydrogeological methods, and long-term water balances. In order to obtain the most reliable results, discharge estimates made by any modeling approaches or marine investigations should be checked against other methods [*Dzhamalov*, 1996].

It is important for estimating nutrient discharges by groundwater into the surface water bodies to evaluate the groundwater capture zone at near-shore zones. *Taniguchi et al.* [1999c] analyzed the groundwater seepage rate into Lake Biwa, Japan, from the point of view of the groundwater capture zone, which is the area and the depth of contributing groundwater discharge into the surface water body. Transient numerical simulations were made using a two-dimensional unsaturated-saturated model with three-layered sediments. The calculated values agreed well with observed groundwater seepage rates for an aquifer thickness of 110 m. This model also agreed with the capture zone estimated from stable isotope data ($\delta^{18}O$ and deuterium). They concluded that aquifer thickness and hydraulic conductivity values are the most important factors for reliable estimates of groundwater seepage rates using theoretical and numerical analysis.

8. CONCLUSIONS

Our main conclusions from this study are summarized below:

(1) Measurements using "manual seepage meters" show that consistent and reliable results can be obtained if one is aware of and careful to prevent known artifacts. New "automated seepage meters" will help to understand the hydrological and coastal oceanographic processes with longer-term and higher-resolution measurements.

(2) Use of natural geochemical or geophysical tracers provides means to scale up to a regional basis of SGD estimates, because the coastal water column tends to integrate natural tracers coming into the system via groundwater pathways. In order to standardize assessment methodologies, intercomparisons between seepage meters and natural geochemical/geophysical tracers are needed to evaluate the variability and deviation of SGD assessments.

(3) Both water balance and hydrograph separation methods may be used for basin-scale estimation of groundwater discharge rates. Unfortunately, the uncertainties in the parameters used in the calculations are often on the same order as the discharge being evaluated.

(4) Theoretically, groundwater discharge from a homogeneous unconfined aquifer into salt or fresh water bodies decreases with distance from the coast. However, some field observations have shown that this is not always the case due to aquifer heterogeneity. Hydrogeological models that account for heterogeneous aquifers on a continental scale are needed to better characterize SGD into the coastal zone.

(5) Numerical methods for estimating SGD depend on estimates of hydraulic conductivity which likely vary over several orders of magnitude within relatively short distances. Estimated groundwater seepage rates also depend on aquifer thickness, another difficult parameter to determine precisely.

Acknowledgments

The authors wish to acknowledge the great help and encouragement from Elizabeth Gross of SCOR (Scientific Committee on Oceanic Research) and Chris Crossland of LOICZ (Land-Ocean Interaction in the Coastal Zone). We also acknowledge Dr. Buddemeier of the University of Kansas, for his insightful comments to improve this manuscript. The first author is also grateful to the Department of Oceanography, Florida State University, and CSIRO Land and Water, Australia, for providing him an opportunity to visit and complete this review work. This work is financially supported in part by SCOR/LOICZ, a grant from the Office of Natural Research (N00014-00-0175) to Florida State University, and a Scientific Research Grant (No.11874064 and No.13304036) from Ministry of Education, Sciences and Culture, Japan. The authors acknowledge to the reviewers and editors for improvement of this manuscript.

REFERENCES

Allen, A.D., 1976. Outline of the hydrogeology of the superficial formations of the Swan Coastal Plain. Western Australia Geol. Surv. Ann. Rep. 31-42.

Anderson, M.P., and Chen, X., 1993. Long and short term transience in a groundwater/lake system in Wisconsin, U.S.A. J. Hydrol. 145, 1-18.

Banks, W. S., Paylor R. L. and Hughes, W. B., 1996. Using thermal-infrared imagery to delineate ground-water discharge. Groundwater 34, 434-443.

Barwell, V.K. and Lee, D.R., 1981. Determination of horizontal-to-vertical hydraulic conductivity ratios from seepage measurements on lake beds. Water Resour. Res. 17, 565-570.

Belanger, T.V. and Montgomery, M.E., 1992. Seepage meter errors. Limnol. Oceanogr. 37, 1787-1795.

Bogle, F.R. and Loy, K., 1995. The application of thermal infrared thermography in the identification of submerged spring in Chickamaura Reservoir, Hamilton County, Tennessee. Karst Geohzards; Proceedings of the fifth multidisciplinary conference on sinkholes and the engineering and environmental impacts of Karst (581p), 415-424.

Bokuniewicz, H. J., 1992. Analytical descriptions of subaqueous groundwater seepage. Estuaries 15, 458-464.

Boldovski, N.V., 1996. Groundwater flow in the coastal zone of the east Sikhote-Alin' volcanogenic belt. Proc. of Int. Symp. on Groundwater Discharge in the Coastal Zone, Land-Ocean Interactions in the Coastal Zone (LOICZ), Moscow, July 6-10, 8-15.

Bouwer, H., 1978. Groundwater Hydrology. McGraw-Hill, New York, 480p.

Boyle, D.R., 1994. Design of a seepage meter for measuring groundwater fluxes in the

nonlittoral zones of lakes – Evaluation in a boreal forest lake. Limnol. Oceanogr. 39, 670-681.
Boyle, J. M., and Saleem, Z. A., 1979. Determination of recharge rate using temperature depth profiles in wells. Water Resour. Res. 15, 1616-1622.
Bredehoeft, J. D., and Papadopulos, I. S., 1965. Rates of vertical groundwater movement estimated from the Earth's thermal profile. Water Resour. Res. 1, 325-328.
Buddemeier, R.W., 1996. Groundwater flux to the ocean: definitions, data, applications, uncertainties. Groundwater discharge in the coastal zone, in LOICZ IGBP, 16-21, edited by R.W. Buddemeier, pp. 179, LOICZ, Texel, Netherlands, Russian Academy of Sciences, Moscow.
Bugna, G.C., Chanton, J.P., Cable, J.E., Burnett W.C., and Cable, P.H., 1996. The importance of groundwater discharge to the methane budgets of nearshore and continental shelf waters of the northeastern Gulf Mexico, Geochimica et Cosmochimica Acta, 60, 4735-4746.
Burnett, W.C., Cowart J.B., and Deetae, S., 1990. Radium in the Suwannee River and Estuary: Spring and river input to the Gulf of Mexico. Biogeochemistry 10, 237-255.
Burnett, W.C., Cable, J.E., Corbett D.R., and Chanton, J.P., 1996. Tracing groundwater flow into surface waters using natural ^{222}Rn. Proc. of Int. Symp. on Groundwater Discharge in the Coastal Zone, Land-Ocean Interactions in the Coastal Zone (LOICZ), Moscow, July 6-10, 22-28.
Burnett, W.C., 1999. Offshore springs and seeps are focus of working group. EOS 80, 13-15.
Burnett, W.C., Taniguchi M., and Oberdorfer, J.A., 2001. Assessment of submarine groundwater discharge into the coastal zone. J. Sea Res. 46(2), 109-116.
Cable, J.E., Bugna, G.C., Burnett W.C., and Chanton, J.P., 1996a. Application of ^{222}Rn and CH_4 for assessment of groundwater discharge to the coastal ocean. Limnol. Oceanogr. 41, 1347-1353.
Cable, J.E., Burnett, W.C., Chanton J.P., and Weatherly, G.L., 1996b, Estimating groundwater discharge into the northeastern Gulf of Mexico using radon-222. Earth and Planetary Sci. Lett. 144, 591-604.
Cable, J.E., Burnett W.C., and Chanton, J.P., 1997. Magnitude and variations of groundwater seepage along a Florida marine shoreline. Biogeochemistry 38, 189-205.
Carr, M.R. and Winter, T.C., 1980. An annotated bibliography of devices developed for direct measurement of seepage,.USGS Open-File Report 80-344, Denver, CO., 38p.
Cartwright, K, 1970. Groundwater discharge in the Illinois Basin as suggested by temperature anomalies. Water Resour. Res. 6, 912-918.
Cartwright, K, 1979. Measurement of fluid velocity using temperature profiles: Experimental verification. J. Hydrol. 43, 185-194.
Cartwright, K, Hunt, C.S., Hughes, G.M., and Brower, R.D., 1979. Hydraulic potential in Lake Michigan bottom sediments. J. Hydrol. 43, 67-78.
Cherkauer, D.S. and McBride, J.M., 1988. A remotely operated seepage meter for use in large lakes and rivers. Ground Water 26, 165-171.
Cherkauer, D.S. and McKereghan, P.F., 1991. Ground-water discharge to lakes: Focusing in embayments. Ground Water 29, 72-80.
Cherkauer, D.S. and Taylor, R.W., 1990. The spatially continuous determination of groundwater flow to surface water bodies: Application to the connecting channels between lakes Huron and Erie. J. Hydrol. 114, 349-369.
Constantz, J., Thomas, C.L., and Zellweger, G., 1994. Influence of diurnal variations in

stream temperature on streamflow loss and groundwater recharge. Water Resour. Res. 30, 3253-3264.
Corbett, D.R., 1999. Tracing Groundwater Flow Into Surface Waters By Application Of Natural And Artifical Tracers. Ph.D. Dissertation, Florida State University, 292 pp.
Cornett, R. J., Risto, B. A., and Lee, D. R., 1989. Measuring groundwater transport through lake sediments by advection and diffusion. Water Res. Res. 25, 1815-1823.
Davis, E.E., Horel, G.C., MacDonald, R.D., Villinger, H., Burnett, R.H., and Li, H., 1991. Pore pressures and permeabilities measured in marine sediments with a tethered probe. J. Geophys. Res. 96(B4), 5975-5984.
Domenico, P. A., and Palciauskas, V. V., 1973. Theoretical analysis of forced convective heat transfer in regional ground-water flow. Geol. Soc. Am. Bull. 84, 3803-3813.
Dzhamalov, R.G., 1996. Methodological approaches to regional assessment of groundwater discharge into the seas, Groundwater discharge in the coastal zone. In LOICZ IGBP, 44-47, edited by R.W. Buddemeier, pp. 179, LOICZ, Texel, Netherlands, Russian Academy of Sciences, Moscow.
Ellins, K.K., Roman-Mas, A., and Lee, R., 1990. Using Rn-222 to examine ground water/surface discharge interaction in the Rio Grande de Manati, Puerto Rico. J. Hydrol. 115, 319-341.
Fairbridge, R. W., 1996. The encyclopedia of oceanography. Reinhold Publishing Corporation, New York, pp. 878-883.
Fellows, C.R. and Brezonik, P.L., 1980. Seepage flow into Florida lakes. Water Resour. Bull. 16, 635-641.
Fischer, W. A., Moxham, R. M., Polcyn, F., and Landies, G. H., 1964. Infrared surveys of Hawaiian volcanoes. Science 146, 733-742.
Freeze R.A. and Cherry, J.A., 1979. Groundwater. Prentice Hall, Englewood Cliff, 604p.
Fritz, P., Cherry, J.A., Weyer, K.U., and Sklash, M., 1976. Storm runoff analyses using environmental isotopes and major ions. In "Interpretation of Environmental Isotope and Hydrochemical Data in Groundwater Hydrology", IAEA, Vienna, 111-130.
Fukuo, Y., 1986. Studies on groundwater seepage in the bottom of Lake Biwa. Report for Environmental Sciences by the Ministry of Education, Science and Culture, Japan, B289-R-12-2, 1-23.
Fukuo, Y. and Kaihotsu, I., 1988. A theoretical analysis of seepage flow of the confined groundwater into the lake bottom with a gentle slope. Water Resour. Res. 24, 1949-1953.
Garrels, R. M. and MacKenzie, F.T., 1971. Evolution of sedimentary rocks. Norton & Co., New York., 397p.
Granier, A., 1985. Une nouvelle methode pour la measure du flux de seve brute dans tronc desarbres. Ann. Sci. For. 42, 81-88.
Harvey, F.E., Lee, D.R., Rudolph, D.L., and Frape, S.K., 1997. Locating groundwater discharge in large lakes using bottom sediment electrical conductivity mapping. Water Resour. Res. 33, 2609-2615.
Hussain, N., Church, T.M., and Kim, G., 1999. Use of ^{222}Rn and ^{226}Ra to trace groundwater discharge into Chesapeake Bay. Marine Chemistry 65, 127-134.
International Atomic Energy Agency, 1979. Isotope in lake studies. Proc. Advisory Group Meet., Vienna, 1977, Vienna.
Israelsen, O. W. and Reeve, R. C., 1944. Canal lining experiments in the delta area, Utah. Utah Agr. Exp. Sta. Tech. Bull. No.313, 52pp.

Jessop, A. M., and Vigrass, I. W., 1989. Geothermal measurements in a deep well at Regina, Saskatchewan. J. Volcanol. Geotherm. Res. 37, 151-166.

Katz, B.G., Lee, T.M., Plummer, L.N., and Busenberg, E., 1995. Chemical evolution of groundwater near a sinkhole lake, northern Florida; 1. Flow patterns, age of groundwater, and influence of lake water leakage. Water Resour. Res. 31, 1549-1564.

Katz, B.G., Coplen, T.B., Bullen, T.D., and Davis, J.H., 1997. Use of chemical and isotopic tracers to characterize the interactions between ground water and surface water in Mantled Karst. Ground Water 35, 1014-1028.

Kohout, F.A., 1966. Submarine springs: A neglected phenomenon of coastal hydrology. Hydrology 26, 391-413.

Kontar, E.A. and Zektser, I.S., 1999. Submarine discharge and its effect on oceanic processes in the coastal zone. Water Resources 26, 512.

Krabbenhoft, D.P. and Anderson, M.P., 1986. Use of a numerical groundwater flow model for hypothesis testing. Ground Water 24, 49-55.

Krabbenhoft, D.P., Bowser, C.J., Anderson, M.P., and Valley, J.W., 1990a. Estimating groundwater exchange with lakes 1. The stable isotopes mass balance method. Water Resour. Res. 26, 2445-2453.

Krabbenhoft, D.P., Bowser, C.J., Anderson, M.P., and Valley, J.W., 1990b. Estimating groundwater exchange with lakes 2. Calibration of a three-dimensional, solute transport model to a stable isotope plume. Water Resour. Res. 26, 2455-2462.

Krupa, S.L., Belanger, T.V., Heck, H.H., Brok, J.T., and Jones, B.J., 1998. Krupaseep – the next generation seepage meter. J. Coastal Res. 25, 210-213.

Lapham, W.W., 1989. Use of temperature profiles beneath streams to determine rates of vertical ground-water flow and vertical hydraulic conductivity. U. S. Geol. Surv. Water Supply Pap. 2337, 35p.

Lee, D.R., 1977. A device for measuring seepage flux in lakes and estuaries. Limnol. Oceanogr. 22, 140-147.

Lee, D. and Harvey, F.E., 1996. Installing piezometers in deepwater sediments. Water Resour. Res. 32, 1113-1117.

Libelo, E. L. and MacIntyre, W. G., 1994. Effects of surface-water movement on seepage-meter measurements of flow through the sediment-water interface. Hydrogeol. J. 2, 49-54.

Linderfelt, W.R., and Turner, J.V., 2001. Interaction between shallow groundwater, saline surface water and nutrient discharge in a seasonal estuary: The Swan River System. Special Swan River edition of Hydrol. Process. (in press).

Lu, N. and Ge, S., 1996. Effect of horizontal heat and fluid flow on the vertical temperature distribution in a semiconfining layer. Water Resour. Res. 32, 1449-1453.

Martin, J.B., Cable, J.E., Swarzenski, P.W., Lindenberg, M., and Hartl, K., 2000. Coastal groundwater discharge to the Indian River Lagoon - physical measurements and water sources. Geological Society of America – Southeastern Section, Charleston, South Carolina, 22-24 March.

Martin, J.B., and Gordon, S.L., 2000. Surface and ground water mixing, flow paths, and temporal variations in chemical compositions of karst springs, In Sasowsky, I.D. and Wicks, C.M. (Eds.), Groundwater Flow and Contaminant Transport in Carbonate Aquifers, A.A. Balkema, Rotterdam, (in press).

McBride, M.S. and Pfannkuch, H.O., 1975. The distribution of seepage within lakebed. J Res.

U. S. Geol. Surv. 3, 505-512.
McDonald, M.G. and Harbaugh, A.W., 1984. A modular three-dimensional finite-difference groundwater flow model. U. S. Geol. Surv.
Moore, W. S., 1996. Large groundwater inputs to coastal waters revealed by ^{226}Ra enrichments. Nature 380, 612-614.
Moore, W. S., and Shaw, T. J., 1998. Chemical signals from submarine fluid advection onto the continental shelf. J. Geophys. Res. - Oceans 103, 21543-21552.
Muir, K. S., 1968. Groundwater reconnaissance of the Santa Barbara – Montecito Area, Santa Barbara County, California. U. S. Geol. Surv. Water Supply Pap. 1859-A, 28p.
Nield, S.P., Townley, L.R., and Barr, A.D., 1994. A framework for quantitative analysis of surface water – groundwater interaction: Flow geometry in a vertical section. Water Resour. Res. 30, 2461-2475.
Oberdorfer, J.A., 1996. Numerical modeling of coastal discharge: predicting the effects of climate change. Groundwater discharge in the coastal zone, in LOICZ IGBP, 69-75, edited by R.W. Buddemeier, pp. 179, LOICZ, Texel, Netherlands, Russian Academy of Sciences, Moscow.
Oberdorfer, J. A., Valentino, M. A., and Smith, S. V., 1990. Groundwater contribution to the nutrient budget of Tomales Bay, California. Biogeochem. 10, 199-216.
Paulsen, R.J., Smith, C.F., and Wong, T., 1997. Development and evaluation of an ultrasonic groundwater seepage meter. Geology of Long Island and Metropolitan New York, Program with Abstracts, Long Island Geologists, State University of New York, Stony Brook, NY, 88-97.
Pluhowski, E. J. and Kantrowitz, I. H., 1964. Hydrology of the Babylon – Islip Area, Suffolk County, Long Island, New York. U.S. Geol. Surv. Water Supply Pap., 1768, 128p.
Rama and Moore, W.S., 1996. Using the radium quartet for evaluating groundwater input and water exchange in salt marshes. GCA, 60 (23), 4245-4252.
Robinson, M. A., 1996. A finite Element Model of Submarine Ground Water Discharge to Tidal Estuarine Waters. PhD dissertation, Virginia Polytechnic Institute.
Roxburgh, I. S., 1985. Thermal infrared detection of submarine spring associated with the Plymouth Limestone. Hydrol. Sci., J. 30, 185-196.
Sakura, Y., 1977. The method for estimations of the groundwater velocity from temperature distributions: On the groundwater around Sapporo, Japan. Stud. Water Temp. 21, 2-14.
Sayles, F.L. and Dickinson, W.H., 1990. The seep meter: a benthic chamber for the sampling and analysis of low velocity hydrothermal vents. Deep-Sea Res. 88, 1-13.
Schultheiss, P.J. and McPhail, S.D., 1986. Direct indication of pore-water advection from pore pressure measurements in Maderia Abyssal Plain sediments. Nature 320, 348-350.
Sellinger, 1995. Groundwater flux into a portion of eastern Lake Michigan. J. Great Lakes Res. 21, 53-63.
Shaw, R.D. and Prepas, E.E., 1989. Anomalous, short-term influx of water into seepage meters. Limnol. Oceanogr. 34, 1343-1351.
Shaw, R.D. and Prepas, E.E., 1990a. Groundwater-lake interactions: I. Accuracy of seepage meter estimations of lake seepage. J. Hydrol. 119, 105-120.
Shaw, R.D. and Prepas, E.E., 1990b. Groundwater-lake interactions: II. Nearshore seepage patterns and the contribution of ground water to lakes in central Alberta. J. Hydrol. 119, 121-136.
Shaw, R.D., Shaw, J.F.H., Fricker, H. and Prepas, E.E., 1990. An integrated approach to

quantify groundwater transport phosphorus to Narrow Lake, Alberta. Limnol. Oceanogr. 35, 870-886.
Shedlock, R. J., Wilcox, D. A., Thompson, T. A., and Cohen, D. A., 1993. Interactions between groundwater and wetlands southern shore of Lake Michigan, USA. J. Hydrol. 141, 127-155.
Shimada, J., Kayane, I., Shimano, Y., and Taniguchi, M., 1993. Use of several tracers to detect the surface-subsurface water interaction in an alluvial fan. IAHS Publ. 215, 263-274.
Sillman, S.E., and Booth, D.F., 1993. Analysis of time-series measurements of sediment temperature for identification of gaining vs. losing portions of Juday Creek, Indiana. J. Hydrol. 146, 131-148.
Smith, L. and Chapman, D.S., 1983. On the thermal effects of groundwater flow, 1, Regional scale system. J. Geophys. Res. 88, 593-608.
Smith, A.J. and Turner, J.V., 2001. Density-dependent surface water-groundwater interaction and nutrient discharge: The Swan-Canning River and estuary system, Western Australia. Special Swan River edition of Hydrol. Process., (in press).
Stallman, R.W., 1963. Computation of groundwater velocity from temperature data. U. S. Geol. Surv. Water Supply Pap, 1544-H, 36-46.
Stallman, R.W., 1965. Steady one-dimensional fluid flow in a semi-infinite porous medium with sinusoidal surface temperature. J. Geophys. Res. 70, 2821-2827.
Stringfield, V. T. and LeGrand, H. E., 1969. Relation of sea water to fresh water in carbonate rocks in coastal areas, with special reference to Florida, U.S.A. J. Hydrol. 9, 387-404.
Stringfield, V. T. and LeGrand, H. E., 1971. Effects of karst features on circulation of water in carbonate rocks in coastal areas. J. Hydrol. 14, 139-157.
Stauffer, R.E., 1985. Use of solute tracers released by weathering to estimate groundwater inflow to a seepage lake. Environ. Sci. Technol., 19, 405-411.
Suzuki, S., 1960. Percolation measurements based on heat flow through soil with special reference to paddy fields. J. Geophys. Res. 65, 2883-2885.
Tanaka, T. and Ono, T., 1998. Contribution of soil water and its flow path to stormflow generation in a forested headwater catchment in central Japan. IAHS Publ. 248, 181-188.
Taniguchi, M., 1993. Evaluation of vertical groundwater fluxes and thermal properties of aquifers based on transient temperature-depth profiles. Water Resour. Res. 29, 2021-2026
Taniguchi, M. and Fukuo, Y., 1993. Continuous measurements of ground-water seepage using an automatic seepage meter. Ground Water 31, 675-679.
Taniguchi, M., 1994. Estimated recharge rates from groundwater temperatures in Nara basin, Japan. Appl. Hydrogeol. 2, 7-13.
Taniguchi, M., 1995. Change in groundwater seepage rate into Lake Biwa, Japan, Jpn. J. Limnol. 56, 261-267.
Taniguchi, M. and Fukuo, Y., 1996. An effect of seiche on groundwater seepage rate into Lake Biwa, Japan. Water Resour. Res. 32, 333-338.
Taniguchi, M., Sakura, Y., and Ishii, T., 1998. Estimations of saltwater-fresh water interfaces and groundwater discharge rates in coastal zones from borehole temperature data. Proceeding of Japanese Association of Groundwater Hydrology Meeting, Tokyo, October 1998, 86-89.
Taniguchi, M., Shimada, J., Tanaka, T., Kayane, I., Sakura, Y., Shimano, Y., Dapaah-Siakwan, S., and Kawashima, S., 1999a. Disturbances of temperature-depth profiles due to surface

climate change and subsurface water flow: 1. An effect of linear increase in surface temperature caused by global warming and urbanization in the Tokyo metropolitan area, Japan. Water Resour. Res. 35, 1507-1517.

Taniguchi, M., Williamson, D. R., and Peck, A. J., 1999b. Disturbances of temperature-depth profiles due to surface climate change and subsurface water flow: 2. An effect of step increase in surface temperature caused by forest clearing in southwest Western Australia. Water Resour. Res. 35, 1519-1529.

Taniguchi, M., Inouchi, K., Tase, N., and Shimada, J., 1999c. Combination of tracer and numerical simulations to evaluate the groundwater capture zone, IAHS Publ., 258, 207-213.

Taniguchi, M., 2000. Evaluation of the saltwater-groundwater interface from borehole temperature in a coastal region. Geophys. Res. Lett. 27, 713-716.

Taniguchi, M. and Iwakawa, H., 2001. Measurements of submarine groundwater discharge rates by a continuous heat – type automated seepage meter in Osaka Bay, Japan. J. of Groundwater Hydrol. 42(4), 271-277.

Taniguchi, M., Burnett, W.C., Cable, J.E., and Turner, J.V., 2002. Investigation of submarine groundwater discharge. Hydrol. Process. (in press).

Tsunogai, U., Ishibashi, J., Wakita, H., and Gamo, T., 1999. Methane-rich plumes in the Suruga Trough (Japan) and their carbon isotopic characterization. Earth and Planetary Sci. Lett. 160, 97-105.

Turner, J.V. and MacPherson, D.K., 1990. Mechanisms affecting streamflow and stream water quality: an approach via stable isotope, hydrogeochemical, and time series analysis. Water Resour. Res. 26, 3005-3019.

Turner, J.V. and Barnes, C.J., 1998. Modeling of isotope and hydrogeochemical responses in catchment hydrology. In: Kendall, C. and McDonnell, J. J., Eds. Isotope tracers in catchment hydrology. Amsterdam, Elsevier, 723-760.

Vanek, V., 1993. Groundwater regime of a tidally influenced coastal pond. J. Hydrol. 151, 317-342.

Williams, J.B. and Pinder III, J.E., 1990. Ground water flow and runoff in a coastal plain stream. Water Resour. Bull. 26, 343-352.

Winter, T.C., 1976. Numerical simulation analysis of the interaction of lakes and groundwater. U.S. Geol. Survey Prof. Paper 1001, 45p.

Winter, T.C., 1983. The interaction of lakes with variably saturated porous media. Water Resour. Res. 19, 1203-1218.

Winter, T.C., 1986. Effect of ground-water recharge on configuration of the water table beneath sand dunes and on seepage in lakes in the sand hills of Nebraska, U.S.A. J. Hydrol. 86, 221-237.

Woessner, W. and Sullivan, K.E., 1984. Results of seepage meter and mini-piezometer study, Lake Mead, Nevada. Ground Water 22, 561-568.

Younger, P.L., 1996. Submarine groundwater discharge. Nature 382, 121-122.

Zektser, I.S. and Dzhamalov, R.G., 1981. Groundwater discharge to the Pacific Ocean. Hydrol. Sci. Bull. 26, 271-279.

Zektser, I.S., Dzhamalov, R.G., and Safronova, T. I., 1983. Role of submarine groundwater discharge in the water balance of Australia. IAHS-AISH Publ. 1, 209-219.

Zektser, I. S., Ivanov, V. A., and Meskheteli, A. V., 1973. The problem of direct groundwater discharge to the seas. J. Hydrol. 20, 1-36.

Zektser, I. S. and Loaiciga, H.A., 1993. Groundwater fluxes in the global hydrologic cycle: past, present and future. J. Hydrol. 144, 405-427.

Edited by M. Taniguchi, K. Wang and T. Gamo

Radon Tracing of Submarine Groundwater Discharge in Coastal Environments

William C. Burnett[a*], Jaye E. Cable[b], and D. Reide Corbett[c]

[a]Department of Oceanography, Florida State University, Tallahassee, FL 32308, USA

[b]Department of Oceanography and Coastal Sciences, Louisiana State University, Baton Rouge, LA 70803, USA

[c]Department of Geology, East Carolina University, Greenville, NC 27858, USA

Direct discharge of groundwater into the coastal zone may be an important material flux pathway from land to sea in some areas. Until relatively recently, it has been largely ignored because of the difficulty in assessing its magnitude. While measurement problems persist, there is a growing recognition that groundwater flow into the sea is important.

We review here an approach, using a simple one-dimensional model, to measure submarine groundwater discharge (SGD) via use of ^{222}Rn as a natural tracer. Radon has certain advantages over other potential geochemical tracers of groundwater discharge. Typically, it is greatly enriched in groundwater compared to seawater, it can be measured at very low concentrations, and is completely conservative. On the other hand, as a gas it is subject to losses at the air-sea interface which may limit its use in shallow water environments.

We conclude that radon tracing is an excellent qualitative tool for identifying areas of spring or seepage inputs in most coastal environments. It can also be a good quantitative tool in shallow marine environments characterized by large amounts of SGD (including re-circulated seawater). The approach is particularly sensitive for inner shelf environments when a strong pycnocline is present as this greatly inhibits radon loss.

1. INTRODUCTION

Although not as obvious as river discharge, continental groundwaters also discharge directly into the sea. Like surface water, groundwater flows down-gradient. Therefore, groundwater flows directly into the ocean wherever a coastal aquifer is connected to the sea. Furthermore, artesian aquifers can extend for considerable distances from shore, discharging groundwaters to the ocean at their points of outcrop. In some cases, these deeper aquifers may have fractures or other breaches in the overlying confining layers, allowing groundwater to flow into the sea. Although such submarine springs and seeps have been known for many years (e.g., written accounts exist from at least the Roman period), these features have traditionally been perceived as hydrologic "curiosities" rather than objects for serious scientific investigation (Kohout, 1966).

This perception is changing. Within the last decade or two recognition has emerged, at

*Corresponding author: Tel: 850-644-6703; Fax: 850-644-2581; email: wburnett@mailer.fsu.edu

least in some cases, that submarine groundwater discharge (SGD) may be both volumetrically and chemically important (Johannes, 1980; Moore, 1999). Estimates of global SGD fluxes vary widely, some estimates are as high as 10% of the river flow while most are considerably lower. Although SGD may not play a significant role in the global water balance, there are reasons to believe that the geochemical cycles of some major and minor elements may be strongly influenced either by the direct discharge of fresh groundwater into the sea or by chemical reactions that occur during the recirculation of seawater through a coastal aquifer system (Zekster et al., 1973; Cathles et al., 1987). In addition, it is now recognized that groundwater discharge may be an important pathway for diffuse pollution to the coastal zone where coastal aquifers have become contaminated by septic systems or other pollution sources (Buddemeier, 1996a).

For many years, investigations of the offshore discharge of groundwater were largely confined to water resource related issues. At least two reasons explain why scientific studies have developed so slowly in this field. First, the SGD process is inherently patchy with significant variability in both time and space, thus making the process very difficult to measure and discouraging serious study. Second, SGD is a process that occurs across a land-sea interface dividing scientific disciplines as well as environments. Unfortunately, there are distinct cultural and structural differences that separate terrestrial and marine scientists (Buddemeier, 1996b). Literally, hydrologists and coastal oceanographers are looking at the same problem from different ends.

Perhaps one of the most promising approaches for regional-scale assessments of SGD is the use of geochemical tracers. The coastal water column tends to integrate natural tracers coming into the system via groundwater pathways. Thus, smaller-scale variations, which would not be of interest for regional studies, are smoothed out. The small scale variability found in many coastal systems has been one of the serious drawbacks concerning the use of seepage meters, a device that provides direct measurements of SGD (Lee, 1977; Burnett, 1999). Many coastal aquifer systems and overlying sediments are by nature very heterogeneous. Thus, the only way to compensate for this natural variability is by making many measurements. Obviously, this restricts the practical coverage of an investigation.

Over the past few years, several studies have employed the use of the natural U decay-series nuclides including ^{226}Ra, short-lived radium isotopes, and ^{222}Rn to assess groundwater inputs to the ocean (Moore, 1996; Rama and Moore, 1996; Cable et al., 1996a/b; Burnett et al., 1996; Moore and Shaw, 1998; Corbett et al., 1999; Hussain et al., 1999; Corbett et al., 2000). Ideally, natural geochemical tracers should be greatly enriched in groundwater relative to coastal waters, be conservative, and be easy to measure. While radium isotopes and radon meet these criteria fairly well, there are certainly other possibilities which may be exploited for groundwater discharge studies. Tritium and helium isotopes would seem particularly attractive in this context although they would not fit the "easy to measure" guideline very well. In actual fact, measurement advances to extend the resolution of all potential groundwater tracers in coastal waters, where concentrations are typically very low, would greatly assist future survey studies.

In applying geochemical tracing techniques, several parameters must be assessed or defined, including boundary conditions (i.e., area, volume), water and constituent sources and sinks (including processes such as adsorption/desorption), residence times of the surface water body, and end-member concentrations of the tracer. Sources or end-members may include ocean water, river water, groundwater, precipitation, in situ production, horizontal water column transport, sediment mixing and resuspension, and diffusion from bottom sediments. Sinks may include in situ decay, horizontal water column transport, horizontal or vertical eddy diffusivity, and atmospheric evasion. Through simple mass balances or box models incorporating both sediment advection and water column transport, the geochemical approach can be quite useful in assessing submarine groundwater discharge.

We present here a description of the ^{222}Rn tracing approach for estimating SGD in the coastal zone. Our intention is to describe how the procedure is applied, point out the various assumptions, explore the sensitivities of various factors relevant to quantifying SGD, and recognize the preferred conditions for application of this method. Two "generic" case studies will illustrate how the approach can be applied in a shallow, well-mixed coastal zone and a stratified system in an inner continental shelf environment. We have attempted to follow the lead of Victor Frankenstein and provide a record of how we use the approach because, like his grandson Frederick, we feel that "...it could work (Brooks, 1974)."

2. SGD ASSESSMENT VIA Rn-222

2.1 Basic Concept & Approach

Basically, an assessment of SGD via radon tracing involves 4 steps: (1) measurement of the water column inventory of ^{222}Rn; (2) an accounting of any ^{222}Rn inputs and outputs to the study area by other processes; (3) a calculation of the total input flux of ^{222}Rn to balance the measured inventories (together with any estimated losses); and (4) a calculation, using estimated fluid concentrations for ^{222}Rn and an advection-diffusion model, of the advective transport required to account for the estimated total input flux. A schematic radon mass balance for a coastal environment is shown in **Figure 1**.

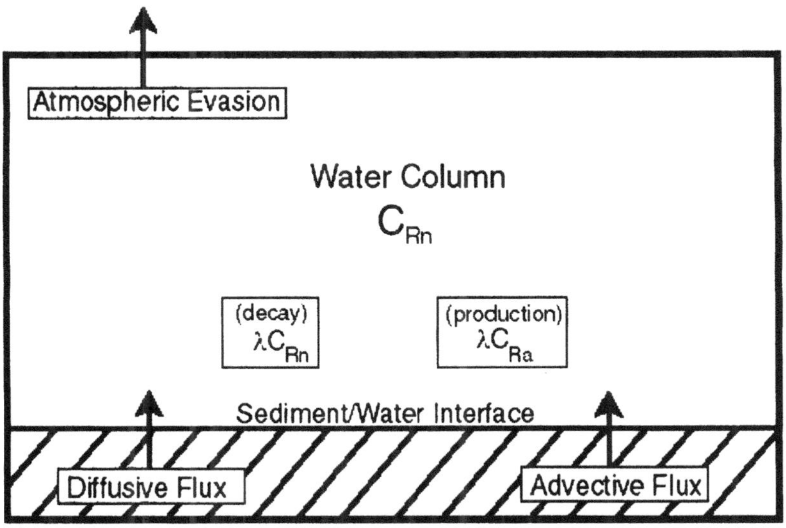

Figure 1 Sources and sinks for ^{222}Rn in a well-mixed coastal environment displayed as a one-dimensional model. The advective flux includes groundwater-driven circulation and physical mixing.

Measurement of the radon concentrations in the water column may be accomplished by standard oceanographic sampling and analysis techniques for measurement of ^{222}Rn taking the special care required for trace gas sampling (Broecker, 1965; Broecker et al., 1968; Mathieu et al., 1988). The data reported in this paper are based on samples collected by a submersible pump into evacuated 4-liter glass bottles. These bottles were then attached to an extraction line and high-purity helium was used to sparge the radon which is collected on a cold trap at liquid nitrogen temperatures. The radon is then transferred to alpha scintillation cells for measurement. Recently, a more automated system has been developed in order to increase the sampling resolution and efficiency of the process (Burnett et al., 2001). An analysis of ^{226}Ra, the parent nuclide of ^{222}Rn, is recommended when possible in order to correct for the "supported" activity of ^{222}Rn, i.e., the amount due to ingrowth from the ^{226}Ra dissolved in the water column. In addition, ^{226}Ra itself may be a powerful groundwater tracer in some environments (Moore, 1996). Once the concentrations have been determined, ideally as a complete profile through the water column, the inventory is calculated by integrating the excess radon concentrations over water depth.

Since radon occurs virtually everywhere in the environment, an accounting of the input and loss terms for the area in question is an important part of this approach (as it would be for any natural tracer). Additional inputs include ingrowth from the ^{226}Ra dissolved in the water column (note that this can be ignored if one uses only "excess" ^{222}Rn, the activity above that which can be supported by ^{226}Ra dissolved in the water) and any benthic inputs from diffusion or via physical mixing (bioturbation, sediment resuspension). An important loss term, at least in a shallow water environment, is loss to the atmosphere. Water mass movement can, of course, transport radon-rich or radon-depleted waters into an area of interest. These can be accounted for if one has sufficient information beyond the specific area being investigated. For the purposes of this discussion, we will assume that horizontal gradients are small and use a one-dimensional model approach to calculate inputs and outputs.

Once one has calculated the inventory, it is relatively simple to estimate the total flux of radon required to support that which has been measured and estimated to have been lost from the system:

$$J = \frac{I}{\left\{ \left(1 - e^{-\lambda t}\right) / \lambda \right\}} \tag{1}$$

where J = the total flux of ^{222}Rn (dpm·m^{-2}·d^{-1}); I = inventory (dpm·m^{-2}), and λ = the decay constant of ^{222}Rn (0.181 d^{-1}). At high values of t (several half-lives of ^{222}Rn), this equation reduces to the inventory divided by the 5.5-day mean life [I/(1/λ)] or simply the inventory multiplied by the decay constant, Iλ. This calculation thus assumes a steady-state situation on a time scale of weeks. This condition has been observed in some coastal environments we have investigated (Corbett et al., 1999) while strong tidal variations have been seen elsewhere (Burnett et al., 2002). The main loss from the measured ^{222}Rn inventory will typically be that due to atmospheric evasion (see section 3.4).

With this estimate of the required total benthic flux of radon, calculations can be made of the advective component required by using an advection-diffusion equation, first presented by Craig (1969) to model radiocarbon in the deep sea and later by Cable et al. (1996a) for SGD estimates:

$$\frac{dC}{dt} = K_z \frac{\partial^2 C}{\partial z^2} + \omega \frac{\partial C}{\partial z} + P + \lambda C \tag{2}$$

where C is the radon concentration (activity) in the wet sediments; z is depth positive downwards; K_Z is the vertical diffusivity; $\frac{\partial^2 C}{\partial z^2}$ and $\frac{\partial C}{\partial z}$ are the ^{222}Rn concentration gradients across the sediment-water interface for diffusion and advection, respectively; ω is the vertical advective velocity; P is the production of ^{222}Rn in pore fluids which is due to recoil after production by ^{226}Ra decay in mineral grains ($P = \lambda C_{eq}$, where C_{eq} is the activity of ^{222}Rn in equilibrium with ^{226}Ra in wet sediment determined experimentally, $dpm \cdot m^{-3}$ wet sediment); and λC is radioactive decay of ^{222}Rn. In this situation, K_Z is set equivalent to D_S, the effective wet sediment diffusion coefficient which is corrected for temperature and sediment tortuosity. Peng et al. (1974) provides an expression for the temperature dependency of molecular diffusion of radon through water, D_o ($-\log D_o = (^{980}/_{T}) + 1.59$) and a correction for sediment tortuosity can be made by multiplying by the sediment porosity (ϕ), i.e., D_S ϕD_O (Ullman and Aller, 1981). For example, the ^{222}Rn molecular diffusion coefficient ($D_O = 1.14 \times 10^{-5}$ $cm^2 \cdot sec^{-1}$ at 18 C; Rona, 1917) becomes 6.0×10^{-6} $cm^2 \cdot sec^{-1}$ for a sediment porosity of 0.42 and a typical summer bottom water temperature of 28.5 °C, the conditions for the inner shelf region of the Gulf of Mexico studied by Cable et al. (1996a). Advection, ω, and radioactive decay, λ, represent losses from the sediments and are thus defined as negative terms. The solution to Eq. 2 may be represented by the following:

$$C = \frac{(C_o - C_{eq})\left(e^{\frac{z}{2z^*}}\right)\sinh\left(\frac{A(z_{eq} - z)}{2z^*}\right)}{\sinh\left(\frac{Az_{eq}}{2z^*}\right)} \tag{3}$$

where C_o is the radon activity ($dpm \cdot m^{-3}$) in the water just overlying the sediment-water interface, multiplied by the sediment porosity to obtain a value corresponding to the ^{222}Rn in wet sediment ($dpm \cdot m^{-3}$); z_{eq} is a depth in the sediments much deeper than the depth where C_{eq} initially occurs; z^* is a one-dimensional mixing parameter described by $^{Ds}/_{\omega}$; and $A = [1 + 4z^* (^{\lambda}/_{\omega})]^{0.5}$, which includes radioactive decay and advection (Craig, 1969). When advection of fluids through sediments is considered, information regarding the radon concentration associated with the subsurface fluids is necessary to estimate accurately the fluid flux across the sediment-water interface. Thus, the estimate of the extent of the water flux through sediments into the overlying water depends critically upon the evaluation of the ^{222}Rn activity in these fluids. If SGD is thought to occur mainly via slow seepage through sediments, a process typically measured at rates on the order of $cm \cdot d^{-1}$, then a reasonable estimate of the fluid radon concentration may be made from the sediment equilibration approach (see below) or from pore water measurements. On the other hand, if more rapid entry points to the sea floor are present, such as submarine springs, then radon activities in the discharging fluids would more likely be similar to those measured in groundwaters from the coastal aquifer. Some knowledge of the pathways of any discharging fluids is thus very helpful for success of this approach.

2.2 One-Dimensional SGD Model

In order to assess groundwater inputs via radon, a simple one-dimensional model may be applied for a steady-state system (Corbett et al., 2000):

$$J + \lambda C_{Ra} z - J_{atm} - \lambda C_{Rn} z = 0 \tag{4}$$

where J represents the total flux of ^{222}Rn from the sea floor to the overlying water column (dpm·m^{-2}·d^{-1}); C is the activity of radium and radon (so λC_{Ra} and λC_{Rn} accounts for production and decay of radon in the water column, dpm·m^{-3}·d^{-1}), respectively; z is the depth of the water column at each site (m); and J_{atm} is the flux of radon from the sea surface to the atmosphere (dpm·m^{-2}·d^{-1}). Horizontal transport of radon is not considered in this approach. We have found that the horizontal gradients are typically small in the coastal environments we have examined in Florida such as Florida Bay (Corbett et al., 1999) and the inner shelf of the northeastern Gulf of Mexico (Cable et al., 1996a). If one is considering a large enough scale, horizontal gradients will clearly become important and mixing must be taken into account.

Sources of radon across the sediment/water interface include both diffusive and advective fluxes. All sediments contain trace amounts of uranium and daughter products including ^{226}Ra, the progenitor of ^{222}Rn. In addition, the short half-life of ^{222}Rn ensures that a steady supply of radon activity will be present from production from ^{226}Ra decay. Thus, diffusion of ^{222}Rn will certainly add some amount of unsupported radon to the overlying waters due to its relatively high mobility in the environment as a noble gas. While diffusive inputs are relatively straight-forward to assess (see below), release of radon into bottom waters via sediment resuspension or other mixing processes remains a very difficult parameter to quantify. In the treatment here, we will assume that such inputs are minor, as they appeared to be in the environments we have investigated around Florida. We acknowledge, however, that neglecting inputs via bioturbation and other forms of physical mixing could result in serious omissions in some situations.

2.3 Input of Radon via Diffusion

There are several methods which have been applied for assessing the diffusive flux of any component across the sediment-water interface: (1) mass balance calculations; (2) measurement of gradients in the sediment and/or water column; and (3) in situ benthic chambers. In addition, Martens et al. (1980) developed a depth-independent mathematical approach specifically for the determination of diffusive fluxes of ^{222}Rn, J_D:

$$J_D = (\lambda D_s)^{1/2}(C_{eq} - C_o) \tag{5}$$

Determination of C_{eq} may be done in the laboratory by equilibration of a measured amount of wet sediment with the ambient water from the environment of interest. We have performed these measurements by placing 70-100 g wet sediment together with 200-300 mL water in stoppered Erlenmeyer flasks equipped with a radon bubbler, equilibrating the ^{222}Rn in water with the sediment by placing the assembly on a shaker table for several weeks, and then analyzing the amount of ^{222}Rn contained in the water via standard radon emanation techniques. We calculate sediment porosity and bulk density from measurements of the percent water in the sediment and independent determinations of the sediment grain density.

In order to validate this approach experimentally, Corbett et al. (1998) estimated ^{222}Rn diffusive fluxes from fresh water lake sediments in a number of ways including the sediment equilibration technique described above, pore water profiles, and by direct measurement in the laboratory of the increase of radon in a water column overlying a simulated sediment bed (**Fig. 2**). The laboratory experiment was conducted by collection of large-volume sediment samples which were then slurried and poured into long (182 cm) 14-cm diameter cast acrylic tubes. The final experiments consisted of an approximate 37-cm sediment bed with an overlying water column of about 120 cm which was monitored for build up of radon over time after any initial radon was sparged out of the system. Radon flux estimates using these three different approaches at two sites in the lake agreed to within 10-15%. The simulated core experiments (similar in principle to a benthic chamber) resulted in the highest values while estimates of diffusive fluxes based on integrating the "missing" ^{222}Rn from the pore water profiles were the lowest.

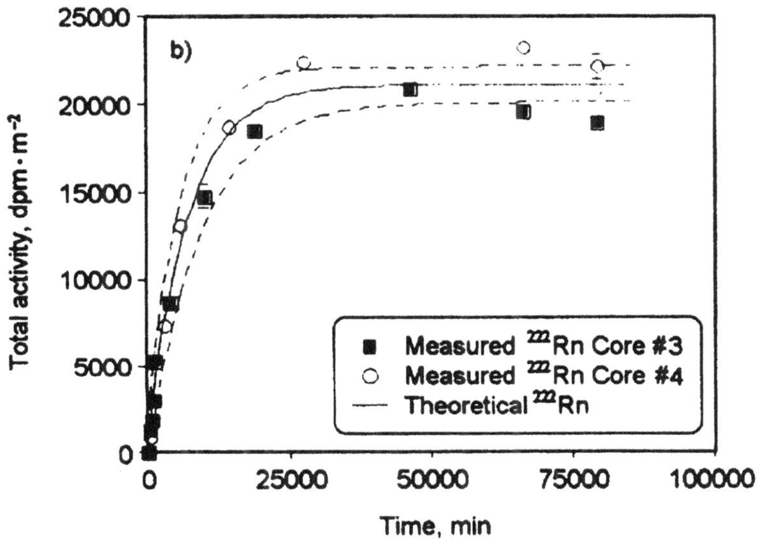

Figure 2 Observed increase in ^{222}Rn inventory in water overlying simulated sediment beds from a freshwater lake. The solid line represents the theoretical development in time of ^{222}Rn with a diffusive flux of 3800±190 dpm·m^{-2}·d^{-1} (Corbett et al., 1998).

Radon transport in soils has been studied extensively over the last few decades because of human health interests involving radon in indoor air and the recognition that soils are typically the main source term (Burnett et al., 1995). Studies have shown that diffusion from soils is dependent upon many factors including radium content, soil moisture, temperature, grain-size distribution, and the location of radium atoms (see Nazaroff et al., 1988 and references therein). Many of these studies have shown that the moisture content of soils, typically quite variable, is particularly important for radon emanation and migration in soils. Sediments, which are always completely saturated, do not have this parameter as a variable. In addition, sediments are typically in a more uniform temperature environment than are soils. We plotted estimated ^{222}Rn diffusive fluxes against measured ^{226}Ra activities in sediments and found a reasonably good relationship (**Fig. 3**). Thus, at least as a first approximation, one can predict the diffusive flux of radon based solely on the ^{226}Ra content of the sediment. The trend shown in Figure 3 is based on a limited data set from just a few environments and so is clearly subject to revision. We recognize that other parameters besides the total amount of ^{226}Ra in the sediment may have significant influence on the diffusion of radon from sediments as well as soils. For example, the specific location of radium atoms (e.g., whether adsorbed onto grain surfaces or lattice bound within sediment minerals) would seem particularly important.

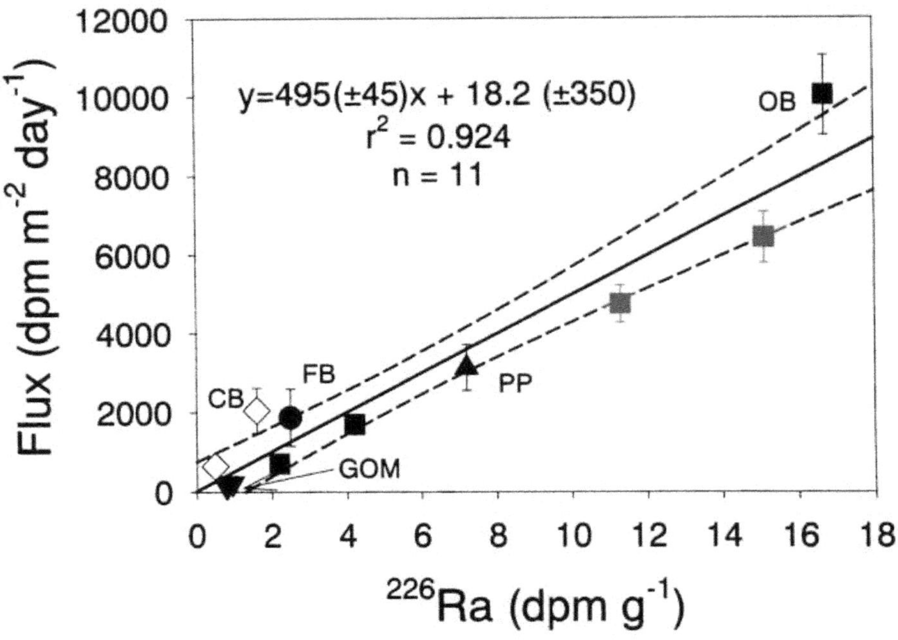

Figure 3 Estimated diffusive flux of ^{222}Rn from sediments in several different environments as a function of the ^{226}Ra activity of those sediments. All sediments are marine except for those from Par Pond, a fresh water lake in South Carolina. GOM (▼) = Gulf of Mexico (Young, 1996); CB (◇) = California Borderland (Berelson et al., 1982); FB (●) = Florida Bay (Corbett et al., 2000); PP (▲) = Par Pond (Corbett et al., 1998); OB (■) = Onslow Bay (Young, 1996).

The diffusion length for radon, defined as $(D_s/\lambda)^{1/2}$ (Nazaroff et al., 1988), ranges from just a few cm in sediments to over 2 m in air. In a sedimentary system with an effective diffusion coefficient of about 6 x 10^{-6} $cm^2 \cdot s^{-1}$, the diffusion length is less than 2 cm. Thus any advective transport at all is likely to dominate material fluxes in such an environment. One may compare advective to diffusive transport directly by use of the following equation:

$$\frac{L_a}{L_d} = v\left(\frac{t}{D_s}\right)^{1/2} \tag{6}$$

where L_a = advective transport distance (cm); L_d = diffusive transport distance (cm); v = advective velocity ($cm \cdot s^{-1}$); t= time (s), and D_s = the effective diffusion coefficient ($cm^2 \cdot s^{-1}$). We plotted this ratio as a function of time to see how long it would take for advective velocities ranging from 1-10 $cm \cdot d^{-1}$ (typical seepage velocities) to equal and then exceed the diffusive transport when D_s = 6 x 10^{-6} $cm^2 \cdot s^{-1}$ (**Fig. 4**). As is clear in the diagram, advection will exceed diffusion after only one day in every case, with transport distances over 14-fold greater after one day and a velocity of 10 $cm \cdot d^{-1}$.

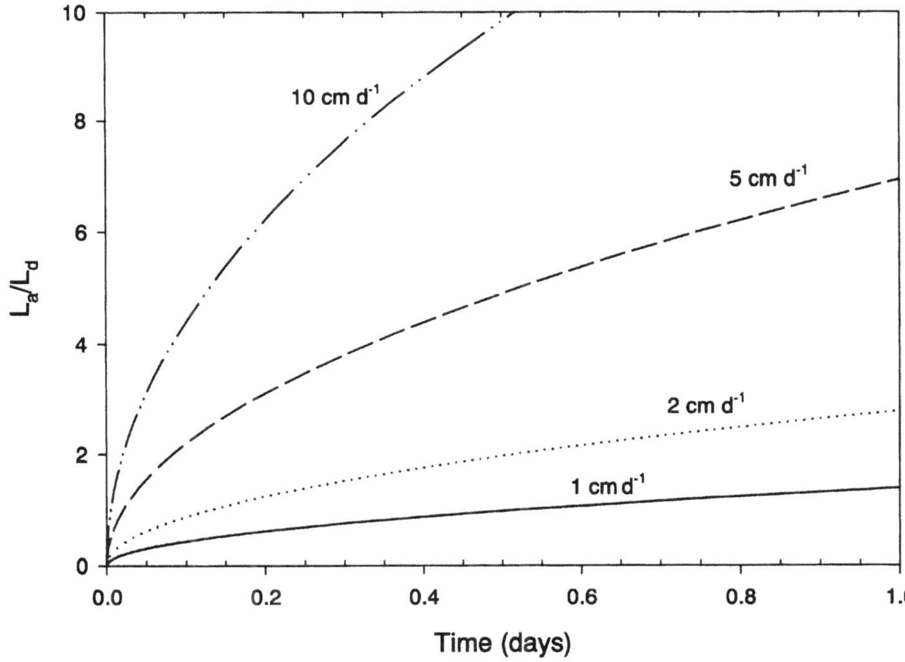

Figure 4 The ratio of the advective transport distance (L_a) to the diffusional transport distance (L_d) as a function of time for 1, 2, 5, and 10 cm·d^{-1} advective velocities and an effective diffusion coefficient for ^{222}Rn in wet sediment of 6.0 x 10^{-6} cm^2·s^{-1}.

2.4 Losses of Radon to the Atmosphere

The total flux across the air-water interface depends on the molecular diffusion produced by the concentration gradient across this interface and turbulent transfer, which is dependent on physical processes, primarily governed by wind speed. The flux (J_{atm}) of radon across the air-water interface can be calculated from an equation presented by MacIntyre et al. (1995):

$$J_{atm} = k(C_w - \alpha C_{atm}) \tag{7}$$

where C_w and C_{atm} are the concentrations of the gas of interest in surface water and air, respectively (dpm·m^{-3}); α is Ostwald's solubility coefficient which defines the distribution of radon between the water and air phases (dimensionless); and *k* is the gas transfer coefficient (m·d^{-1}), a function of the kinematic viscosity, molecular diffusion, and turbulence which is dependent on wind speed (Jahne et al., 1987; Wanninkhof, 1992; MacIntyre et al., 1995; Bugna et al., 1996; Corbett et al., 1997).

The gas transfer coefficient, k, is a function of the physical processes at the air-sea boundary, especially the turbulence and kinematic viscosity of the water (ν), and the molecular diffusion coefficient of the gas (D_m = 1.16 x10^{-5} cm^2·s^{-1} at 20°C for radon). The Schmidt number (Sc) is the ratio of the kinematic viscosity to the molecular diffusion coefficient, i.e., Sc= ν/D_m. Based on a number of field studies, empirical equations which relate k to wind speed have been proposed. Macintyre et al. (1995) present an equation where k represents the piston velocity for a given wind speed normalized to the Schmidt number for CO_2:

$$k(600) = 0.45u_{10}^{1.6}(Sc/600)^{-0.5} \quad (8)$$

where u_{10} is the wind speed at 10 m height above the water surface and Sc for the dissolved gas of interest is divided by 600 to normalize k to CO_2 at 20°C in freshwater. Turner et al. (1996) showed in calculating the gas transfer coefficient for DMS as a function of wind speed, that the (Sc/600) term in Eq 8 should be raised to the power of –0.667 for $u_{10} \le 3.6$ m·s^{-1} and –0.5 for $u_{10} > 3.6$ m·s^{-1}.

Gesell (1983) showed that atmospheric ^{222}Rn activities are typically on the order of 220 to 890 dpm·m^{-3}. In a more maritime environment, perhaps more representative of the coastal systems of interest here, Kim et al. (2000) reported radon-in-air activities ranging from 60-550 dpm·m^{-3} around an area of northern Chesapeake Bay. Using the range for atmospheric ^{222}Rn reported in Kim et al. (2000) and the ^{222}Rn activities typical of coastal waters in the northeastern Gulf of Mexico (Cable et al., 1996b), we have calculated the radon flux to the atmosphere, J_{atm}, at different temperatures and air-sea concentration gradients as a function of wind speed (**Fig. 5**). It is clear from this diagram that both wind speed and temperature are very important in controlling the loss of radon across the sea surface while the concentration of radon in the air has less importance, especially when the radon activity in the water is relatively high (6000 dpm·m^{-3} given in the example). At lower coastal water ^{222}Rn activities (2000·dpm·m^{-3} in Fig. 5b), the difference in the atmospheric concentration becomes noticeable but is still not a major control. Radon fluxes to the atmosphere become very high, relative to typical diffusive fluxes for example, at wind speeds above about 5 m·s^{-1} while it is much less important under about 2 m·s^{-1}.

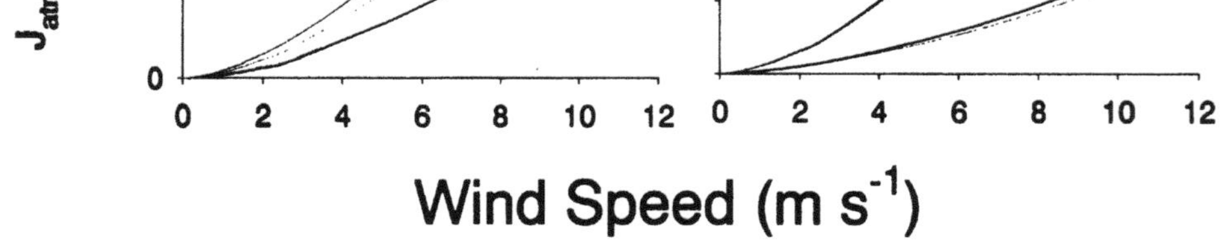

Figure 5 Estimated fluxes of ^{222}Rn across the water-air interface as a function of wind speed as well as: (A) temperature; and (B) water-air concentration gradients. C_w = radon activity in water; C_a = radon activity in air. Radon-in-water concentrations are typical of the coastal areas near the Florida State University Marine Laboratory, FSUML (Young, 1996). Atmospheric concentrations represent the range reported in Kim et al. (2000).

3. EXAMPLE CALCULATION: COASTAL ONE-LAYER ENVIRONMENT

We have constructed a radon balance (**Fig. 6**) to evaluate SGD in a shallow (<2 m) near-shore (<1000 m from shore) environment in the Big Bend region of northwestern Florida near the Florida State University Marine Laboratory (FSUML) based on dozens of individual ^{222}Rn measurements made at 3 seepage meter transects during 1992-94. This area has been shown to have significant seepage with distinct seasonal trends (Bugna et al., 1996; Cable et al, 1996b; Cable et al., 1997). Measurements made during this period included ^{222}Rn and ^{226}Ra in the near-bottom water, sediment ^{226}Ra, physical properties (sediment porosity, bulk density), fluxes of ^{222}Rn via benthic chambers, and seepage rates via standard Lee-type seepage meters (Young, 1996, Cable et al., 1996b).

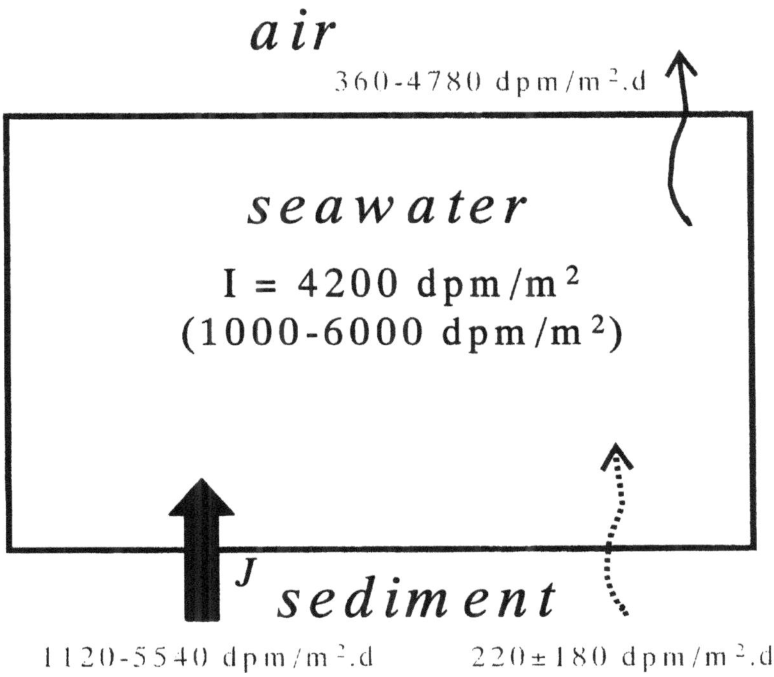

Figure 6 One-layer, one-dimensional radon budget for the shallow coastal area near FSUML, northeastern Gulf of Mexico.

Observed inventories of ^{222}Rn are in the range of 1000 to 6000 dpm·m^{-2} with an average of around 4200 dpm·m^{-2} at the station closest to shore. If there was no loss to the atmosphere (extremely unlikely in this shallow environment), this inventory would require a steady-state benthic flux of about 760 dpm·m^{-2}·d^{-1} to balance loss by radioactive decay. Including atmospheric losses, of course, would require a corresponding higher benthic flux for a mass balance. Estimated diffusive fluxes from the quartz-rich muddy sand sediments in the area are very low, about 100-200 dpm·m^{-2}·d^{-1} based on sediment equilibration measurements, pore water measurements, and extrapolation to zero seepage of a linear trend of ^{222}Rn inventories versus measured seepage rates (Cable et al., 1996b). Thus, there is clearly another source of excess ^{222}Rn in this area and the seepage meter data confirms that groundwater flow is the most likely candidate.

Benthic chambers made from plexiglass with openings for stirring and sampling were deployed at this site to measure radon fluxes. Some ^{222}Rn measurements were made with the chambers in a "seeping" mode, i.e., ports on the top of the chambers were left open so that the flux calculated should represent the total flux, not just the diffusive component, of the input. The results of these measurements showed a range of 4000-7000 dpm·m^{-2}·d^{-1} with a mean value of 5360±1070 dpm·m^{-2}·d^{-1}, 20-30 times higher than the estimated diffusive fluxes. If sufficient data do not exist to make an independent estimate of the radon loss to the atmosphere, we can estimate this loss for a steady-state system by subtracting the flux required to balance the observed inventory from the total flux from the benthic chambers,

$$J_{atm} = J_{total} - \frac{I}{\frac{1}{\lambda}} \quad (9)$$

where J_{total} is the total ^{222}Rn flux as determined via benthic chamber measurements. In our example, the estimated atmospheric flux is thus (5360±1070 dpm·m^{-2}·d^{-1}) – (760 dpm·m^{-2}·d^{-1}) = 4600 dpm·m^{-2}·d^{-1}. This estimated loss seems reasonable as it can easily be accounted for with average winds of 3-4 m·s^{-1} in the study area at 20°C (Fig. 5). Since most of our measurements were made in the summer months when water temperatures approach 30°C, such an atmospheric loss could be accounted for with even gentler winds.

Using the advection-diffusion model (equation #2) and measured sediment and bottom water parameters (**Table 1**: C_o = 4200 dpm·m^{-3}; C_{eq} = 97000 dpm·m^{-3}; ϕ =0.40) from the area,

Table 1. Water column, sediment, and other parameters used for radon tracing calculations to estimate SGD in two areas.

Parameter	*FSUML Shallow Coastal Environment*	*K-Tower Inner Shelf, Pycnocline*
Decay constant for ^{222}Rn, λ	0.181 d^{-1}	
Molecular diffusion coefficient for ^{222}Rn in water, D_o	1.16 x 10^{-5} cm^2·s^{-1}	
Effective wet sediment ^{222}Rn diffusion coefficient, D_s	6.0 x 10^{-6} cm·s^{-1}	
Inventory, I	4200 dpm·m^{-2}	21300 dpm·m^{-2}
Vertical eddy diffusivity, K_v	na	3.4 x 10^{-4} cm^2·s^{-1}
^{222}Rn activity in overlying water, C_o	4200 dpm·m^{-3}	4000 dpm·m^{-3}
Activity of ^{222}Rn in equilibrium with wet sediment, C_{eq}	97000 dpm·m^{-3}	168000 dpm·m^{-3}
Porosity, ϕ	0.40	0.43
Total ^{222}Rn flux measured via seeping chambers, J	5360 dpm·m^{-2}·d^{-1}	2520 dpm·m^{-2}·d^{-1}
Estimated SGD velocity, ω	7 cm·d^{-1}	2.5 cm·d^{-1}

we estimate that an upwelling rate equivalent to 7 $cm \cdot d^{-1}$ at the sediment-water interface is needed to account for the additional radon influx into the area. Most of the actual seepage meter measurements made at the three transects near FSUML were in the range of 4-8 $cm \cdot d^{-1}$ while the mean value of the most shoreward of the stations during the same period as the radon measurements was 8.3±6.5 $cm \cdot d^{-1}$ (n=37). Thus, while the radon approach agrees reasonably well with the actual seepage measurements in this area, the degree of uncertainty is high because of the large range of possible atmospheric losses. The benthic chamber experiments allowed us to constrain this somewhat but these measurements are somewhat difficult to make and are also subject to large variations (about 20% in this case but often larger). Should the diffusive fluxes be greater (because of higher sediment ^{226}Ra, for example), or the seepage rates lower, it would be difficult to extract the necessary information with this approach. On the other hand, the radon tracing approach has the distinct advantage of integrating all input terms and thus providing a means to scale up from a purely local level to a more regional basis.

4. EXAMPLE CALCULATION: INNER SHELF TWO-LAYER ENVIRONMENT

The inner shelf region of the Gulf of Mexico in the Big Bend Region of Florida, south of FSUML, is characterized by a well-mixed water column in the winter with development of a strong pycnocline in the summer (June-August) with estuarine-like circulation (Marmorino, 1983; Cable et al., 1996a). We made repeated visits during 1992-94 for water column sampling in an 620 km^2 area centered around K-Tower, a navigational tower maintained by the U.S. Air Force, about 24 km southeast of FSUML. During this period, 17 complete water column profiles of ^{222}Rn and ^{226}Ra were obtained immediately adjacent to K-Tower during all seasons. Of these 17 profiles, 8 were obtained during summertime conditions when a well-developed pycnocline was present (Young, 1996). These results, together with many other measurements within the entire study area, were used by Cable et al. (1996a) to estimate the total amount of SGD. Their estimate, based on a coupled benthic exchange – water column transport (2-dimensional) model, showed that this area experiences about 180-710 $m^3 \cdot s^{-1}$ of SGD, approaching the average flow (720 $m^3 \cdot s^{-1}$) of the Apalachicola River, the largest river in Florida. We will show here, using a simpler one-dimensional approach, that one can make nearly the same estimates and with much less uncertainty than in the one-layer case considered above where potentially large unaccounted for losses to the atmosphere exist.

When both the total and sub-pycnocline ^{222}Rn inventories of the 1992-94 K-Tower results are plotted (**Fig.** 7), it is clear that considerably more variation occurs in the total inventories than those in the summertime below the pycnocline. During the three summers when these measurements were made, the sub-pycnocline ^{222}Rn inventories had a mean value of 21,300±2100 $dpm \cdot m^{-2}$, or only about a 10% standard deviation, which is about the same as the precision of the radon measurements themselves. The total inventories show dramatic seasonal fluctuations, not surprising in view of the changing oceanographic conditions during the year. Even during the summertime periods, however, the total inventories are more variable than those below the pycnocline. This is likely a result of the surface waters being in contact with the shoreline which is influenced by river inputs and surficial aquifers. Presumably, these surficial aquifers are more sensitive to shorter-term changes in rainfall and other environmental conditions than the deeper Floridan Aquifer (the main fresh water aquifer in the southeastern U.S.) which is the likely source of waters seeping into these offshore coastal waters.

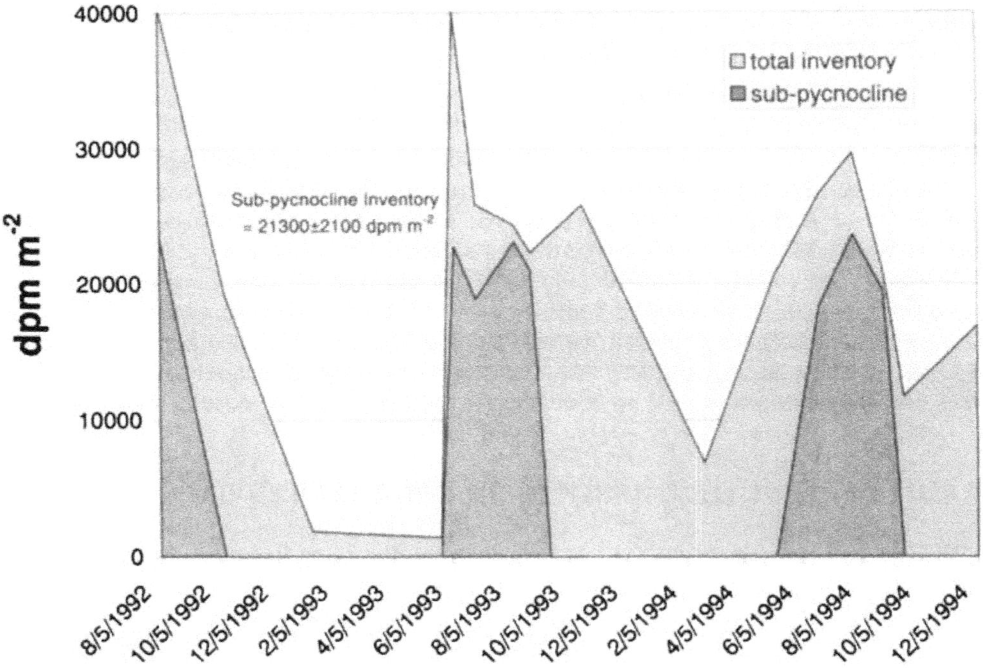

Figure 7 Total and sub-pycnocline inventories of ^{222}Rn in the water column immediately adjacent to K-Tower from August, 1992 to December, 1994. The mean sub-pycnocline inventory for this period was 21300±2100 dpm·m^{-2}·d^{-1} (n=8; Young, 1996).

Mainly because ^{222}Rn loss across the well-developed pycnocline is so slight, the 2-layer model (**Fig. 8**) is greatly simplified relative to the 1-layer near-shore system which is subject to fluctuating losses to the atmosphere. The sediments at K-Tower are a mixture of quartz sand and shell hash, are low in ^{226}Ra, and produce a ^{222}Rn diffusive flux of only about 211±12 dpm·m^{-2}·d^{-1} via the sediment equilibration method (n=4). Sediments from the entire study area reported by Cable et al. (1996a) were similar with an estimated diffusive flux of ^{222}Rn of 230±110 (n=13) dpm·m^{-2}·d^{-1}. The flux of ^{222}Rn across the pycnocline (J_{pyc}) may be estimated by:

$$J_{pyc} = K_v(\frac{dC}{dz}) \quad (10)$$

where (dC/dz) represents the concentration gradient of ^{222}Rn across the pycnocline and K_v is the vertical eddy diffusivity which is a measure of the turbulent transfer of density. The relationship which governs K_v is its inverse dependence on the buoyancy gradient (i.e., density gradient) (Pond and Pickard, 1983; Gargett, 1984). Using an average summertime ^{222}Rn concentration gradient at K-Tower and a value of 3.4(±1.8)x10^{-4} cm^2·s^{-1} for K_v, we estimate that the J_{pyc} is very low at about 4.4 dpm·m^{-2}·d^{-1}. Cable et al. (1996a) derived K_v using a reported oceanic buoyancy flux (N) of $4x10^{-6}$ $cm^2 \cdot s^{-3}$ (Sarmiento et al, 1976) and density gradients $\left[\frac{g}{\rho}\right]\left(\frac{\partial \rho_{pot}}{\delta z}\right)$ estimated from 7 hydrographic stations measured in the summertime in the vicinity of K-Tower. This value for J_{pyc} is extremely low, even compared to the estimated diffusive fluxes for this area. Neglecting this small loss across the pycnocline, we can easily estimate that a total flux of 3860±380 $dpm \cdot m^{-2} \cdot d^{-1}$ would be required to provide a steady-state inventory equivalent to the observed 21,300±2100 $dpm \cdot m^{-2}$. Open benthic chambers deployed at K-Tower produced somewhat lower fluxes of 1700-3360 $dpm \cdot m^{-2} \cdot d^{-1}$ with a mean value of 2520±670 $dpm \cdot m^{-2} \cdot d^{-1}$ (Cable et al., 1996a). Using the measured sediment and ^{222}Rn concentration parameters for this area (Table 1; C_o=4000 $dpm \cdot m^{-3}$, C_{eq}=168000 $dpm \cdot m^{-3}$, ϕ=0.43), we used the advection-diffusion model to estimate that an upwelling rate of 2.5 $cm \cdot d^{-1}$ at the sediment-water interface is necessary to provide the required total estimated flux of 3860 $dpm \cdot m^{-2} \cdot d^{-1}$. If we assume that this estimate is representative of the entire 620 km^2 area studied by Cable et al. (1996a), then a total discharge rate is estimated at 180±15 $m^3 \cdot s^{-1}$, within the same range (but at the lower end) of that estimated by the more complex coupled water transport – benthic exchange approach.

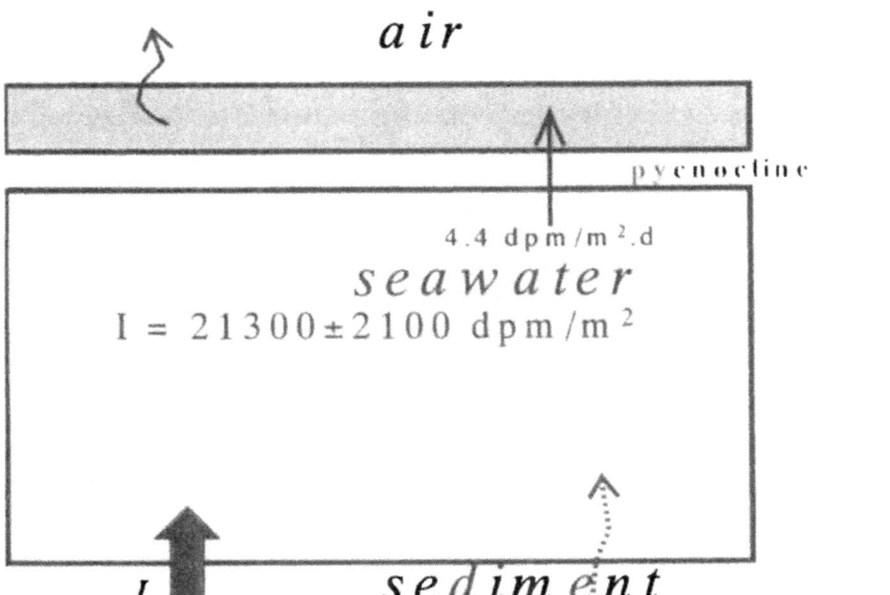

Figure 8 Two-layer, one-dimensional radon budget for the sub-pycnocline waters immediately adjacent to K-Tower on the inner shelf of the northeastern Gulf of Mexico.

6. CONCLUSION

Radon tracing is a powerful tool for qualitative assessment of SGD in coastal environments. In some environments, where SGD is particularly high or where other conditions are favorable (low ^{226}Ra in the sediments, well-developed pycnocline, no significant inputs of radon via sediment mixing) and some additional information is available concerning pathways of groundwater into the area, ^{222}Rn can be used to make good quantitative SGD estimates. Areas we have examined around Florida fit these requirements nicely and we have been able to make discharge estimates consistent with other available information.

As a natural tracer, radon (as all natural tracers) will be present in the coastal environment from other sources which requires an assessment of these inputs as well as sinks. Diffusion from bottom sediments may be estimated reliably in a variety of ways, but benthic inputs as a result of sediment resuspension, biological mixing, or other types of physical mixing are difficult to quantify. Inputs and losses via production from dissolved ^{226}Ra and ^{222}Rn decay may be calculated precisely if the ^{226}Ra is measured as well as ^{222}Rn. Losses to the atmosphere may be estimated from wind speed, temperature, and concentration data. Unfortunately, these estimates are typically going to have large uncertainties because of high variability, especially in wind speeds. Working in calm, uniform conditions reduces these types of uncertainties considerably. Even better, using sub-pycnocline radon inventories would completely eliminate the need to assess the air-sea interaction term in many cases. Benthic chamber experiments, while not absolutely necessary, can add additional information which often helps evaluate ^{222}Rn fluxes.

One of the main advantages of the geochemical tracer approach to estimating SGD is that the water column tends to integrate the measured tracer which may be derived from various seeps, springs, etc. over the region of interest. This has the advantage of smoothing out small-scale variations which tend to be typical of many coastal environments. Thus, one has the opportunity to scale up from the local (meters to kilometer) to the regional (kilometers to tens of kilometers) setting. This is clearly important for assessment of SGD over long shorelines.

Acknowledgments

The authors are grateful to the personnel at the FSU Marine Laboratory for their assistance in our research. Dr. Joselene De Oliveira (IPEN, Sao Paulo, Brazil) asked a sufficient number of difficult questions to inspire the need for this paper. We thank Mel Brooks for providing an interesting, amusing, and useful point of view. This manuscript is a contribution to Working Group 112, "Magnitude of Submarine Groundwater Discharge and its Influence on Coastal Oceanographic Processes" sponsored by the Scientific Committee on Oceanic Research (SCOR) and the Land-Ocean Interactions in the Coastal Zone (LOICZ) project of IGBP. Financial support for the radon tracing investigations has been provided by grants from the National Science Foundation, National Oceanic and Atmospheric Administration, Earthwatch, and currently by the Office of Naval Research (N00014-00-0175).

REFERENCES

Berelson, W., Hammond, D. and Fuller, C., 1982. Radon-222 as a tracer for mixing in the water column and benthic exchange in the southern California borderland. Earth Planet. Sci. Lett., 61, 41-54.

Broecker, W.S., 1965. An application of natural radon to problems in oceanic circulation. In: Proc. Symp. On Diffusion in the Oceans and Freshwaters. Lamont Geological Observatory, New York, 116-145.

Broecker, W.S., Cromwell, J., and Li, Y.-H., 1968. Earth Planet. Sci. Lett., 5, 101.

Brooks, Mel, 1974. "Young Frankenstein." 20th Century Fox.

Buddemeier, R.W. (ed), 1996a. Groundwater Discharge in the Coastal Zone: Proceedings of an International Symposium. LOICZ IGBP. LOICZ/R&S/96-8, iv+179pp. LOICZ, Texel, The Netherlands.

Buddemeier, R.W., 1996b. Groundwater flux to the ocean: definitions, data, applications, uncertainties. In Buddemeier, R.W. (ed), Groundwater Discharge in the Coastal Zone: Proceedings of an International Symposium. LOICZ IGBP. LOICZ/R&S/96-8, iv+179pp. LOICZ, Texel, The Netherlands, 16-21.

Bugna, G.C., Chanton, J.P., Cable, J.E., Burnett, W.C., and Cable, P.H., 1996. The importance of groundwater discharge to the methane budgets of nearshore and continental shelf waters of the northeastern Gulf of Mexico. Geochim. Cosmochim. Acta, 60, 4735-4746.

Burnett, W.C., Cable, P.H., and Chanton, J. P., 1995. A simple passive collector for direct measurement of radon flux from soil. Jour. Radioanalytical Nuclear Chem., 193, 281-290.

Burnett, W.C., 1999. Offshore springs and seeps are focus of working group. EOS, 80, 13-15.

Burnett, W.C., Cable, J.E., Corbett, D.R., and Chanton, J.P., 1996. Tracing groundwater flow into surface waters using natural ^{222}Rn. In Buddemeier, R.W. (ed), Groundwater Discharge in the Coastal Zone: Proceedings of an International Symposium. LOICZ IGBP. LOICZ/R&S/96-8, iv+179pp. LOICZ, Texel, The Netherlands, 22-28.

Burnett, W.C., Kim, G., and Lane-Smith, D., 2001. Use of a continuous radon monitor for assessment of radon in coastal ocean waters. Jour. Radioanal. Nucl. Chem., 249, 167-172.

Burnett, W.C., Lambert, M., Christoff, J., Chanton, J., Taniguchi, M., Moore, W., Smith, L., Krupa, S., Kontar, E., Smith, C., Paulsen, R., and O'Rourke, D., 2002. SCOR/LOICZ group conducts assessment of groundwater discharge to the ocean. EOS, submitted.

Cable, J.E., Burnett, W.C., Chanton, J.P., and Weatherly, G., 1996a. Modeling groundwater flow into the ocean based on ^{222}Rn. Earth Planet. Sci. Lett., 144, 591-604.

Cable, J., Bugna, G., Burnett, W., and Chanton, J., 1996b. Application of ^{222}Rn and CH_4 for assessment of groundwater discharge to the coastal ocean. Limnol. Oceanog., 41, 1347-1353.

Cable, Jaye E., Burnett, W.C., and Chanton., J.P., 1997. Magnitude and variations of groundwater seepage along a Florida marine shoreline, Biogeochemistry, 38, 189-205.

Cathles, L.M. et al., 1987. Fluid circulation in the crust and the global geochemical budget. In: G.B. Munsch (Editor), Second Conference on Scientific Ocean Drilling (COSOD II). European Science Foundation, Strasbourg, France.

Corbett, D.R., Burnett, W.C., and Cable, P.H., 1997. Tracing of groundwater input into Par Pond, Savannah River Site by Rn-222. Jour. Hydrology, 203, 209-227.

Corbett, D.R., Burnett, W.C., Cable, P.H., and Clark, S.B., 1998. A multiple approach to the determination of radon fluxes from sediments. Jour. Radioanalytical & Nuclear Chemistry, 236, 247-252.

Corbett, D.R., Chanton, J., Burnett, W., Dillon, K., Rutkowski, C., and Fourqurean, J., 1999. Patterns of groundwater discharge into Florida Bay. Limnol. Oceanogr., 44, 1045-1055.

Corbett, D.R., Dillon, K., Burnett, W., and Chanton, J., 2000. Estimating the groundwater contribution into Florida Bay via natural tracers ^{222}Rn and CH_4. Limnol. Oceanogr., 47, 1546-1557.

Craig, H., 1969. Abyssal carbon and radiocarbon in the Pacific. J. Geophys. Res., 74, 5491-5506.

Gargett, A. 1984. Vertical eddy diffusivity in the ocean interior, *J. Mar. Res.* 42: 359-393.

Gesell, T.F., 1983. Background atmospheric Rn-222 concentrations outdoors and indoors: A review. Health Phys., 45, 289-302.

Hussain, N., Church, T.M., and Kim, G., 1999. Use of ^{222}Rn and ^{226}Ra to trace groundwater discharge into Chesapeake Bay. Marine Chemistry, 65, 127-134.

Jahne, B., Munnich, K.O., Bosinger, R., Dutzi, A., Huber, W., Libner, P., 1987. On the parameters influencing air-water gas exchange. J. Geophys. Res., 92, 1937-1950.

Johannes, R.E., 1980. The ecological significance of the submarine discharge of groundwater. Marine Ecology - Progress Series, 3, 365-373.

Kim, G., Hussain, N., and Church, T.M., 2000. Excess ^{210}Po in the coastal atmosphere. Tellus, 52B, 74-80.

Kohout, F.A., 1966. Submarine springs: A neglected phenomenon of coastal hydrology. Hydrology, 26, 391-413.

Lee, D.R., 1977. A device for measuring seepage flux in lake and estuaries. Limnol. Oceanogr., 22, 140-147.

Macintyre, S., Wanninkhof, R., and Chanton, J.P., 1995. Trace gas exchange across the air-sea interface in freshwater and coastal marine environments. In: P.A. Matson, R.C. Harriss (eds.) Biogenic Trace Gases: Measuring Emissions from Soil and Water, Blackwell Science Ltd. 52-97.

Marmorino, G., 1983. Variability of current, temperature, and bottom pressure across the West Florida continental shelf, winter 1981-1982. J. Geophys. Res. 88, 4439-4457.

Martens, C.S., Kipphut, G.W., and Klump, J.V., 1980. Sediment-water chemical exchange in the coastal zone traced by in situ radon-222 flux measurements. Science, 208, 285-288.

Mathieu, G., Biscayne, P., Lupton, R., and Hammond, D., 1988. System for measurements of ^{222}Rn at low levels in natural waters. Health Physics, 55, 989-992.

Moore, W.S., 1996. Large groundwater inputs to coastal waters revealed by ^{226}Ra enrichments. Nature, 380, 612-614.

Moore, W.S., 1999. The subterranean estuary: A reaction zone of groundwater and sea water. Marine Chemistry, 1-24.

Moore, W. S. and Shaw, T. J., 1998. Chemical signals from submarine fluid advection onto the continental shelf. J. Geophys. Res. Oceans, 103, 21543-21552.

Nazaroff, W.W., Moed, B.A., and Sextro, R.G., 1988. Soil as a source of indoor radon: generation, migration, and entry. In: Radon and Its Decay Products in Indoor Air (W.W. Nazaroff and A.V. Nero, eds.), John Wiley & Sons, New York, 57-112.

Peng, T.-H., Takahashi, T., and Broecker, W., 1974. Surface radon measurements in the north Pacific station Papa. J. Geophys. Res. 79, 1772-1780.

Pond, S. and Pickard. G., 1983. Introductory Dynamical Oceanography, Pergamon Press, Oxford, 329 pp.

Rama and Moore, W.S., 1996. Using the radium quartet for evaluating groundwater input and water exchange in salt marshes. Geochim. Cosmochim. Acta, 60, 4245-4252.

Rona, E., 1917. Diffusionsgrösse und atomdurchmesser der radiumemanation, Zeitschrift für Physikalische Chemie 92, 213-218, 1917.

Sarmiento, J., Feely, H., Moore, W., Bainbridge, A., Broecker, W., 1976. The relationship between vertical eddy diffusion and buoyancy gradient in the deep sea. Earth Planet. Sci. Lett., 32, 357-370.

Turner, S.M., Malin, G., Nightingale, P.D., Liss, P.S., 1996. Seasonal variation of dimethyl sulphide in the North sea and an assessment of fluxes to the atmosphere. Marine Chemistry, 54, 245-262.

Ullman, W. and Aller, R., 1981. Diffusion coefficients in nearshore marine sediments. Limnol. Oceanogr., 27, 552-556.

Wanninkhof, R., 1992. Relationship between wind speed and gas exchange over the ocean. J. Geophys. Res., 97, 7373-7382.

Young, J.E., 1996. Tracing Groundwater Flow into the Northeastern Gulf of Mexico Using Naturally Occuring Radon-222. Ph.D. Dissertation, Florida State University, 263 pp.

Zektzer, I.S., Ivanov, V.A., and Meskheteli, A.V., 1973. The problem of direct groundwater discharge to the seas. Jour. Hydrology, 20, 1-36.

Land and Marine Hydrogeology
Edited by M. Taniguchi, K. Wang and T. Gamo

The chemical characteristics of submarine groundwater seepage in Toyama Bay, Central Japan

J. Zhang and H. Satake

Faculty of Science, Toyama University
Gofuku 3190, Toyama, 930-8555 Japan

The discharge of freshwater from a continental shelf directly into the marine environment, which is linked to the groundwater system of the land, has historically been ignored as a pathway for water and dissolved materials, both natural and anthropogenic, to the ocean. In Toyama, located in central Japan, the average annual precipitation exceeds the total amount of evaporation and river runoff. It is possible that this difference in water quantity can be attributed to water flowing into a groundwater spring from the continental shelf decades after permeating into the soil of the land. This quantity corresponds to about 30% of the river runoff to Toyama Bay. In other words, a nutrient-rich groundwater spring at the sea floor is a significant source of nutrition, more so than the river water, to the coastal marine ecosystem of Toyama Bay. Here we review several studies of shoal-seepage systems from the sea floor off of the Katagai and Kurobe alluvials, using chemical techniques to clarify the circulation mechanisms and their spatiotemporal changes.

1. INTRODUCTION

There are several important sources for various materials that flow from the land to the ocean, and one of these sources is fresh water. The input and distribution of material to the marine environment by means of riverine inflow, atmospheric precipitation or geothermal input due to the eruption of submarine volcanoes on mid-oceanic ridges have been considered and discussed. However, the direct discharge of freshwater from the continental shelf into the marine environment, which links to the groundwater system of the land, has thus far been ignored as a pathway for water and dissolved materials, both natural and anthropogenic, to the ocean. Submarine springs, well known by fishermen as good fishing spots, input massive amounts of nutrients into coastal marine ecosystems and therefore influence those ecosystems.

By the development of recent research techniques, the presence of submarine spring water was discovered to be a common phenomenon in continental shelf areas around the world. Spring water has been found in the northern and southern American continental shelves off California (Muir, 1986; Oberdorfer et al., 1990), Florida (Back and Hanshaw, 1970; Barraclough and Marsh, 1962; Belanger and Walker, 1990; Belanger et al., 1997; Brooks et al., 1993; Cable et al., 1997; Cao et al., 1999; Chanton et al., 1991; Cooper and Tindall, 1994;

Corbett, 1999), and New York (Bokuniewicz, 1980; Flipse and Bonner, 1985); the islands of the Bahamas (Henderson et al., 1999; Whitaker and Smart, 1990) and Jamaica (D'Elia et al., 1981); Mexico's Yucatan Peninsula (Back and Lesser, 1981; Hanshaw and Back, 1980; Moore et al., 1992; Stoessell et al., 1989); several areas in the Pacific region including Hawaii (Dollar and Arkinson, 1992; Kanehiro and Peterson, 1977; Sansone and Resing, 1995), Guam (Matson, 1993), Australia (Dickson, 1985; Johannes and Hearn, 1985; Wright, 1999), and Japan (Igarashi et al.,, 1995; Komae, 1990; Taniguchi and Fukuo, 1996; Taniguchi et al., 1999; Tsunogai et al., 1996); in France (Cathles, 1987; Marriotti et al., 1988), Greece (Burden, 1964; Stringfield and LeGrand, 1969; Higgins, 1980); and elsewhere. Burnett et al. (2001) concluded that the existences of these submarine groundwater spring in three different types are: (1) nearshore seepage, (2) offshore seepage and (3) submarine spring. Although it has been noticed that some of these springs are large enough to supply the amount of freshwater needed to meet human requirements, even now there is little knowledge of the physics and chemistry of water that seeps from the sea floor. For example, little is known about the flux in the seepage of such water from a continental shelf, the degeneration of the groundwater during the runoff process under the terrestrial soil and marine sediment, or submarine springs' effects on coastal chemical processes or upon marine ecosystems.

Here, to provide a model case of groundwater discharge from a shelf, recent studies of Toyama Bay, located in central Japan, are reviewed in order to clarify the circulation mechanism and the spatiotemporal changes of a shoal seeping system from the sea floor, as determined by chemical techniques.

2. GEOGRAPHIC AND OCEANOGRAPHIC SETTINGS

Toyama Bay is located at the foot of the North Nippon Alps and Tateyama Mountains, as shown in Fig. 1 and Fig. 2. As the maximum water depth exceeds 1200 m in the center, Toyama is one of the three largest and deepest bays in Japan, along with Sagami Bay and Suruga Bay, which are on the other side of the Japanese islands. Since Toyama Bay has been connected geomorphologically through the Toyama submarine canyon to the central Japan Sea, the seawater below a water depth of 300 m, known as Japan Sea Proper Water which flows perpetually into the bay, is low in temperature and rich in nutrients. Since part of the Tsushima Warm Current, which is a tributary of the Kuroshio, flows into Toyama Bay, the bay is blessed with both warm- and cold-water species of fish. Furthermore, several groundwater springs from the continental shelf are reported to be in this region (Fujii, et al., 1986), in addition to the inflow of rivers and the upwelling of deepwater (Japan Sea Proper Water) in the bay's inner part. All of these are important nutrient sources for Toyama Bay, supporting the coastal fishery environment in the Japan Sea. However, there have been few systematic researches from the chemical-oceanographic viewpoint, such as studies of the characteristics of the coastal seawater, the flow processes of the deep water or estimations of residence time.

Holocene-age submerged tree remains were found off the continental shelf of Japan bordering the Japan Sea near Yoshihara, Nyuzen-machi, Toyama Prefecture, adjacent to the Kurobe alluvial (Fig 2., Nasu et al., 1983). The remains, standing tree stumps and roots, are located at depths between 20 and 40 meters, and ages of 8,000 to 10,000 years B.P. have been determined by the ^{14}C method (Konishi, 1983). The distribution of the stumps and the roots suggests that there were probably two spatially separate forests, which were overwhelmed by

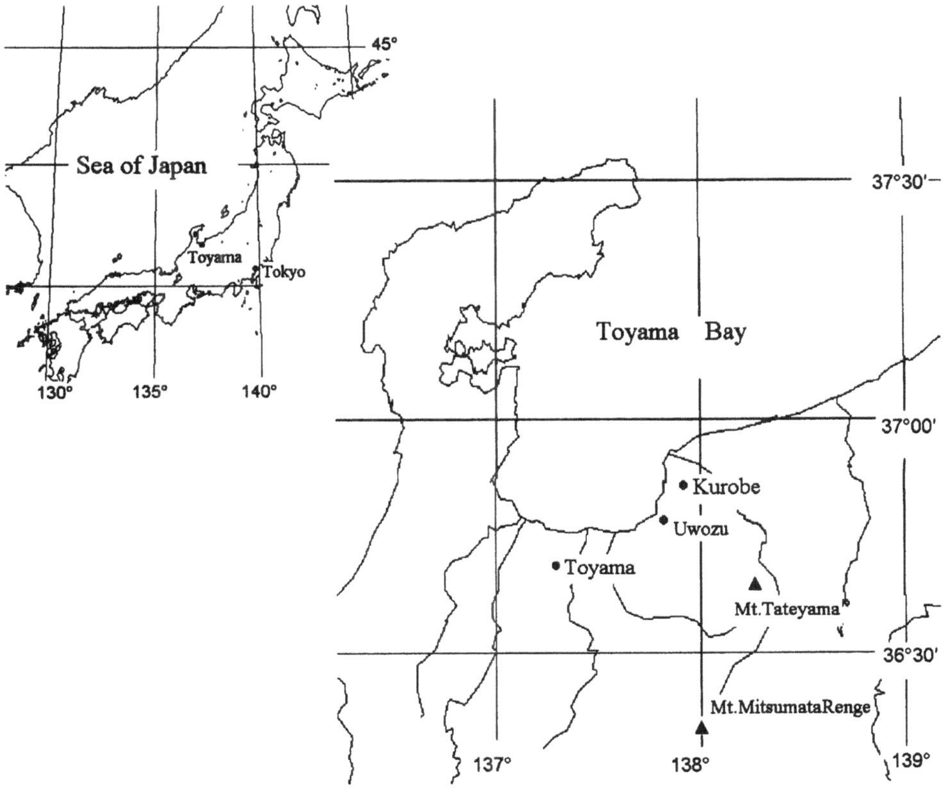

Fig. 1. Location map of Toyama Bay, central Japan.

a rise in sea level and then preserved by flash-flood debris (Fujii et al., 1986). The existing submarine exposures may be due to recent submarine erosion related to the seaward movement of floodwaters that passed from the fan into the upper reaches of the Toyama submarine canyon system.

The sub-aerial Kurobe fan is well watered and has abundant groundwater near the coast. This, together with the location of the submerged forest off Nyuzen and Holocene buried forest in Uwozu described early by Fuji (1971), makes it easy to estimate that the fresh water emerges from the submerged fan and causes a number of submarine springs at the inside and the edge of the Toyama shelf, as reported previously (Fujii, et al., 1986). However, groundwater discharge to the ocean as a supply route of freshwater and various dissolved components has not been considered in this area. We are now conducting geochemical surveys in order to determine the chemical features of the seepage water system in Toyama Bay and to investigate the seasonal variations of these features, while evaluating quantitatively their effect on the coastal ocean environment. The results will be published elsewhere.

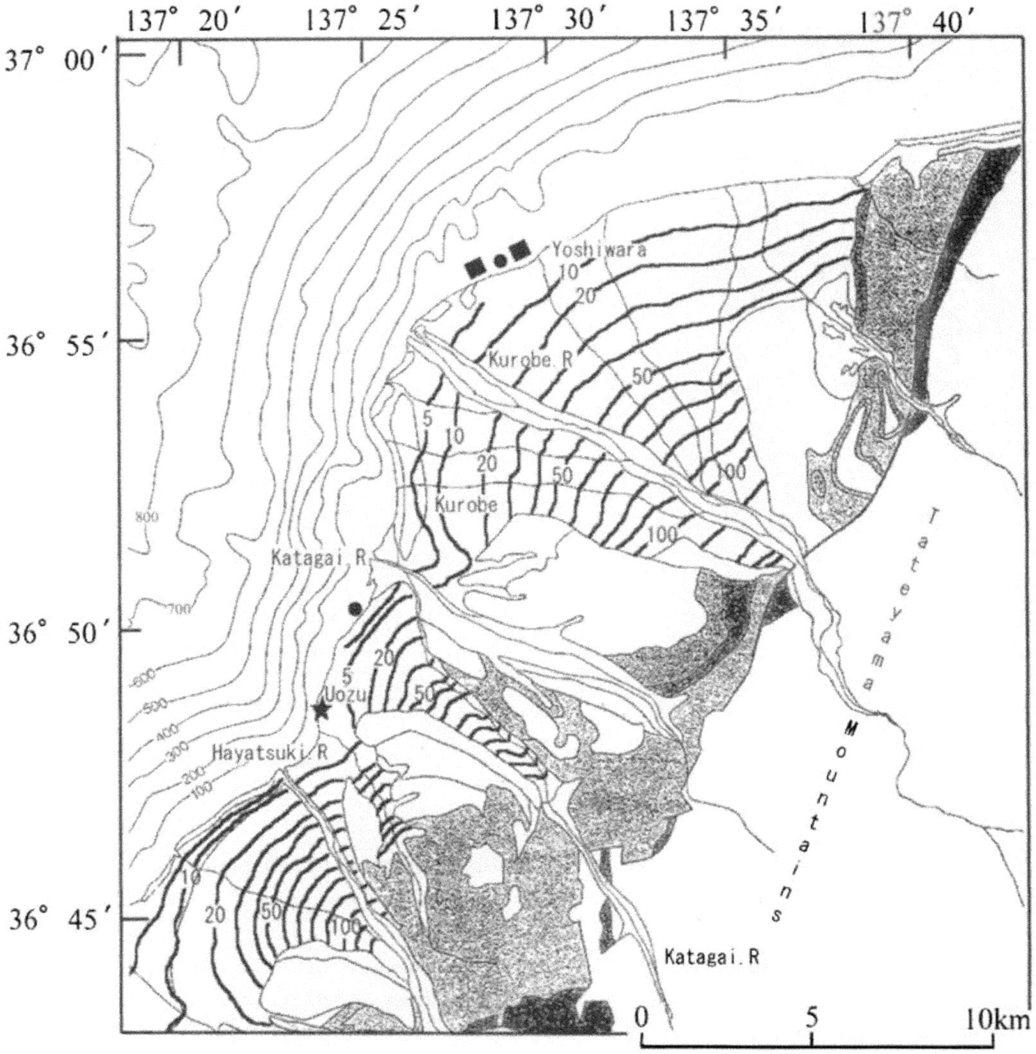

Fig. 2. General settings of the buried forest in Uwozu (★), submerged forests (■) off Kurobe alluvial and submarine groundwater seepage (●) off Uwozu in Toyama Bay, central Japan.

3. SAMPLING METHODS

A seepage site was proposed for off Uozo, where the buried forest had been found and a large number of naturally flowing springs exist on land (Zhang et al., 2001). As a result, the spring position of the groundwater at the sea floor was found, as shown in Fig. 2, by a hearing investigation of the bay's circumference by the fishery association and the Fisheries Experimental Station of Toyama Prefecture. The investigation included the collection of groundwater by diving to the spring point, and a wide survey and seawater sampling of the circumference of the ocean environment using two research boats, the *Tateyama-maru* and *Hayatsuki,* both of the Toyama Prefectural Fisheries Experimental Station.

Low-salinity water determinations, at –27 m near the submerged tree remains off Kurobe estuary, were reported in 1986 (Fig. 3, Fujii et al). These waters were recovered by divers

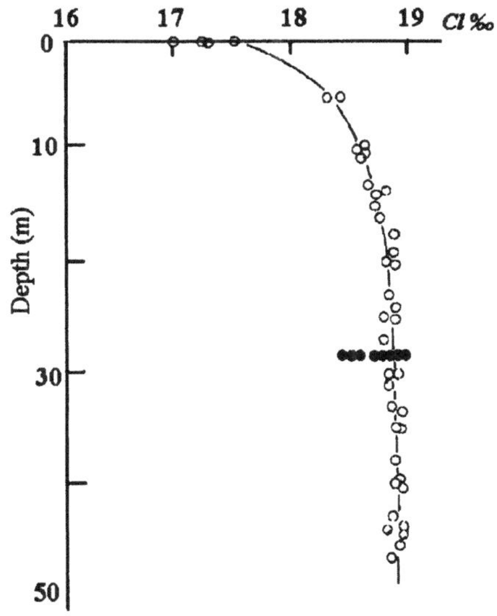

Fig. 3. Salinity profiles of seawater (o) and spring waters (●) off Kurobe. Reproduced from Fujii et al., 1986.

from beneath using syringes, and interpreted as a mix of 5% groundwater upwelling from the sea floor. Since the ionic strength of seawater is extremely high compared with that of groundwater, detailed chemical analysis in these springs was difficult. A new technique was applied to the freshwater sampling at the spring point, using a system that combined hollow and 1- m-long stainless pipes, two three-way cocks and a pre-cleaned plastic tetra bag (Nakata et al., 2001; Tokunaga et al., 2001). This method enabled the collection of a large amount of fresh groundwater without any contamination by seawater. After this collection, the following analyses were carried out: temperature, salinity, dissolved oxygen, hydrogen and oxygen isotope ratio, nutrients, major ion concentration, tritium, rare earth elements and other chemical components. Further analyses were conducted to determine the mechanism by which chemical material is transformed as a result of groundwater seepage from the seafloor and the mechanism by which terrestrial freshwater degenerates during the transfer between the land and ocean under the soil and/or sediment. In addition, a compact-memory type CT (conductivity and temperature) and a TD (temperature and density) sensor (Alec Electronics Ltd.) were each moored at the spring point in order to monitor the seasonal variations of temperature and salinity.

4. CLIMATE AND CHEMICAL CHARACTERISTICS

A recently developed research technique has detected spring groundwater throughout the sea floor, corresponding to the end part of the freshwater subterranean. In addition, the existence of spring groundwater flowing in a shallow groundwater mass has also been discovered where the water rises through the deep mass along the saltwater/freshwater interface. Marui and Yasuhara (1999) classified groundwater that seeps from the sea floor into the following three patterns (Fig. 4):

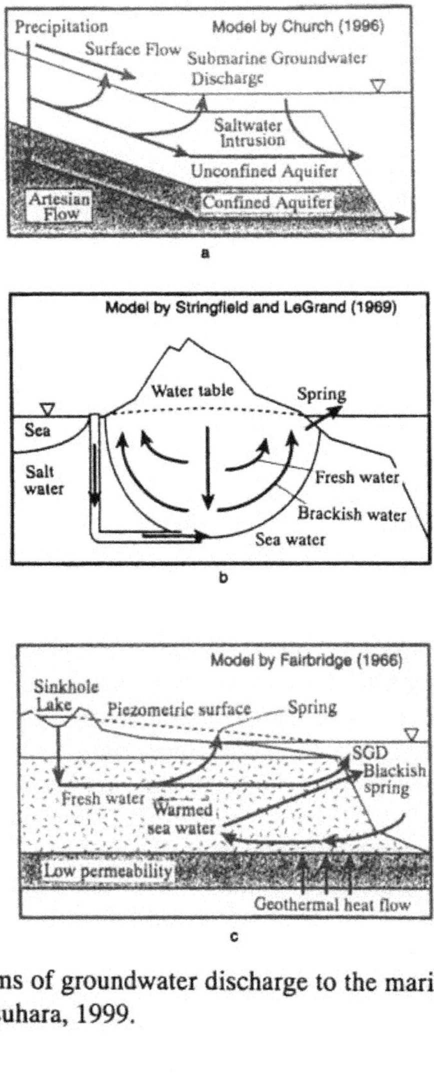

Fig. 4. Pattern diagrams of groundwater discharge to the marine environment. Reprint from Marui and Yasuhara, 1999.

a) The potential type: The flow of the groundwater is controlled by the potential of the spring in terms of conversion points of the geomorphologic gradient.
b) The density type: Different water densities cause an upward flow, while the groundwater permeates toward the deeper water and comes into contact with the seawater.
c) The geothermic type: Heat from deep volcanic sources causes the groundwater to rise.

The seepage water collected off Uwozu is of the potential type. The source of the freshwater is the precipitation in the Toyama region, which is among the highest in Japan. Most cities in the world with the same latitude as Toyama are semi-arid, and have an average annual precipitation of 600 mm. However, the annual precipitation at Toyama is 2296 mm actually in creasing to 4000 mm (Fujii, 2000) in the eastern mountains such as the Mt. Mitsumata-Renge (2,841m) and Mt. Kekatsu (2,414m). The precipitation at Toyama is almost twice as high as that of Tokyo (Fig. 5), which has an annual precipitation of 1405 mm (annual average for 1961 to 1990, according to the Tokyo meteorological observatory). Toyama's high precipitation is caused by high seawater temperature (over 10℃), which is maintained by the warm Tsushima current moving at 2.6 million m^3/sec from the southwest to the northeast along the Japan Sea coast. As a result, winter is wet and warm in the coastal zone of Toyama, because evaporation is rapid in the Japan Sea in winter (Fujii, 2000).

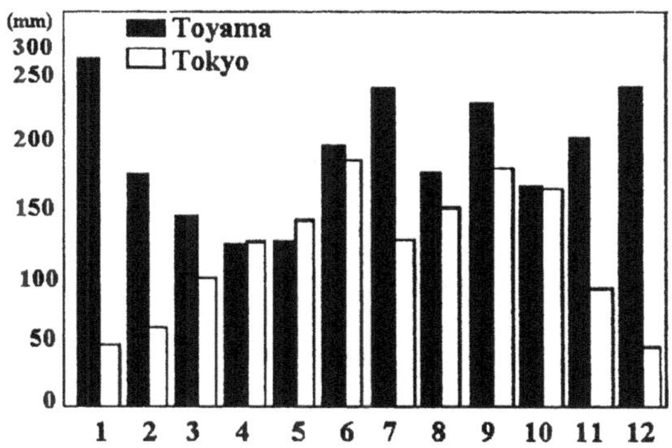

Fig. 5. Comparative monthly precipitation at Toyama and Tokyo.

The stable isotope ratios of hydrogen and oxygen and the major ion component are useful tracers for understanding hydrological features; for example, the origin and catchments of the groundwater, water-rock interaction and deep-water behavior. The groundwater spring off Uwozu is located in the estuaries of both the Hayatsuki River and Katagai River. The major chemical compositions of the spring water reflect the value of these two rivers, indicating that it originates from them (Zhang et al., 2001).

5. OXYGEN AND HYDROGEN ISOTOPIC COMPOSITION

If we plot the relationships between oxygen and hydrogen isotopes of spring water, river water and groundwater (well water), we find that they are distributed between two straight lines with a gradient of 8, but with two different y-intercepts (Fig. 6), as shown in the following equations:

$$\delta D = 8\delta^{18}O + 10 \quad (1)$$

$$\delta D = 8\delta^{18}O + 30 \quad (2)$$

The y-intercept here is called the d-parameter, which indicates the degree of the dynamic fractionation of the steam when it is evaporated from the seawater and diffused in the atmosphere.

In precipitation, the isotope of oxygen and that of hydrogen on a global scale are almost definite and are distributed on the straight line (Craig et al., 1961) of equation (1). However, the precipitation of Toyama is plotted on two straight lines as shown in Fig. 6, with a d-parameter of 10 for summer precipitation and a d-parameter of 30 for the snow in winter (Waseda and Nakai, 1983; Satake et al., 1983). The details are explained as follows.

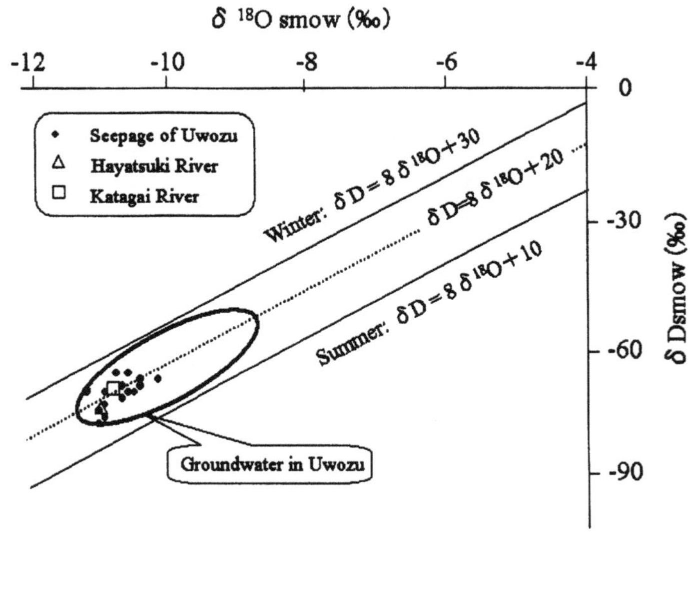

Fig. 6. δD and $\delta^{18}O$ distributions of groundwater, seepage and river waters in the eastern Toyama region.

As vapor forms in the summer, the humid maritime air mass (Ogasawara air mass) forms slowly from Pacific Ocean water, with a low d-parameter of the precipitation. In winter, on the other hand, the rapid evaporation into a dried continental cold air mass (Siberian air mass) from the warm Japan Sea causes a higher d-parameter. As a result, the average d-parameter of the Japanese rainwater gradually rises from about 10 on the Pacific Ocean side to above 20 on the Japan Sea side. Since the geomorphological variation is very complicated in the Toyama region, the local air mass, which is controlled by climate change, varies intensely between summer and winter, causing the d-parameter of 10 in the summer and 30 in the winter. Therefore, as a tracer of the surface water and groundwater, the d-parameter should be an effective index that helps us to understand the catchments, the water recharge source and its flow systems.

Moreover, because of the altitude effect of oxygen and hydrogen in the precipitation as described below, the oxygen and hydrogen isotope compositions also fluctuate according to the average catchment height of the river water and/or groundwater. Mizutani and Satake (1997) published a calculation formula that describes the relationship between the altitude and isotopic composition of the recharge source in the surface water and groundwater of Toyama, using changes in the oxygen and hydrogen isotope compositions and the d-parameter.

$$\delta D = -(0.0193 \pm 0.017)\,H - (48.4 \pm 2.3) \quad (3)$$

$$\delta^{18}O = -(0.00236 \pm 0.00016)\,H - (8.68 \pm 0.22) \quad (4)$$

where H stands for the altitude in meters.

Using these equations, the average recharge altitude of the seepage water was calculated to be about 840 m from its δD and $\delta^{18}O$ values (Zhang, et al., 2001).

The Hayatsuki River and Katagai River, as shown in Fig. 1, whose estuaries are near the groundwater spring area, formed small-scale alluvial fans at the mountains connecting to the coast (Fujii, 1992). Because the edges of the alluvial fans contact the sea directly or are close to the coastline, the groundwater possibly permeates into the soil and springs from the sea floor through the alluvial fans. However, considering that the catchment altitude of the Hayatsuki River is 1200 m and that of the Katagai River is 840 m, the groundwater collected from off Uwozu is originated from precipitation that has fallen on mountains with an average altitude of 850 m (Fig. 7). As a general condition, the $\delta^{18}O$ and δD of the river water are comparatively stable (Mizutani and Oda, 1983; Mizutani et al., 2001), fluctuating minimally throughout the year. This indicates that the precipitation permeates underground and then mixes completely before flowing to the river (Mizutani and Satake, 1997).

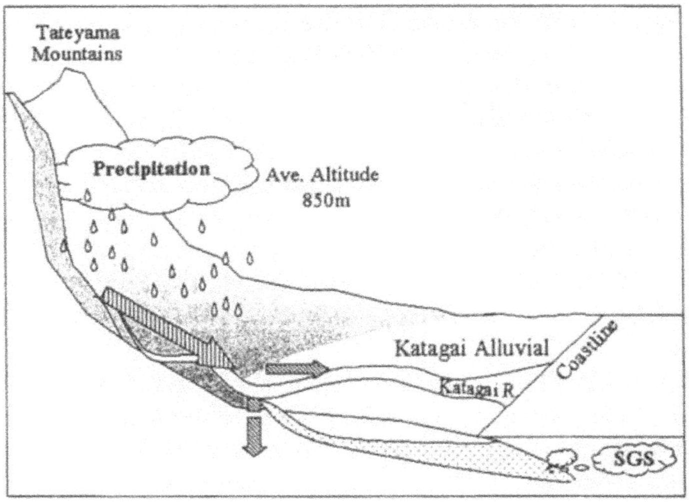

Fig. 7. Model sketch of the submarine groundwater seepage (SGS) off Uwozu.

6. TRITIUM RECORDS

Tritium is also believed to be one of the important indexes for comprehending the flow route and residence time of the river water and groundwater. However, it is necessary to first clarify the background of the tritium concentration in the local surface water (such as precipitation and river water) before investigating the water origin using the tritium composition. The natural tritium concentration rose rapidly in the 1960's due to atomic and hydrogen bomb experiments, reaching 770 T.U. (annual average) in Tokyo in 1963. Since nuclear tests in the atmosphere have stopped, the tritium concentration has gradually decreased. In particular, tritium concentrations in rain and river water were greatly reduced during the 1990s, from 30 T.U. to almost 5 T.U., approaching the pre-nuclear-testing values. Under such circumstances, accurate measurement techniques are necessary in order to calculate the water system division, origin and residence time in the groundwater based on the tritium concentration. Highly accurate data were obtained with an error margin of about ±0.5 T.U. when a low-background type liquid scintillation counter measurement was used, after the electrolytic enrichment of tritium using SPE (solid polymer electrolyte, Saito et al., 1996).

By virtue of the fact that tritium and helium-3 in the groundwater originate from tritium in precipitation, groundwater can be dated using the following equation:

$$N = N_0 \exp(-\lambda t) \tag{5}$$

where λ is decay constant; N is the tritium concentration in the groundwater and N_0 represents the sum of tritium and helium-3 concentrations in the groundwater. Since the tritium concentration of seepage water off Uwozu is 3.9±0.5 (T.U.), the spring groundwater gushing from the sea floor over 10 to 20 years, after having permeated the underground, was distinguishable by comparing it with the historical record of tritium concentrations in Toyama precipitation (Satake and Kanda, 1989; Satake, 2001) and with those for six large local rivers (Satake et al., 1984), based on a piston flow model.

7. IMPACT ON THE COASTAL ENVIRONMENTS

In Toyama, the average annual precipitation (2,296mm) exceeds the total amount (201 x 10^8 m^3/yr) of evaporation and river runoff (Fig. 8, Ito and Fujii, 1993; Fujii, 2000). It is possible that this difference in water quantity is diverted by a groundwater spring from the continental shelf decades after precipitation has permeated into the soil of the land. This quantity corresponds to about 30% of the river runoff to Toyama Bay. In addition, because some of the Tateyama Mountains are 3000 m in height near the Toyama coast, and the basis rock under the Kurobe alluvial fan is found at a water depth of about 600 m, it can be estimated that the pressure is sufficient to push the groundwater seeping from the sea floor up to a water depth of less than 600 m.

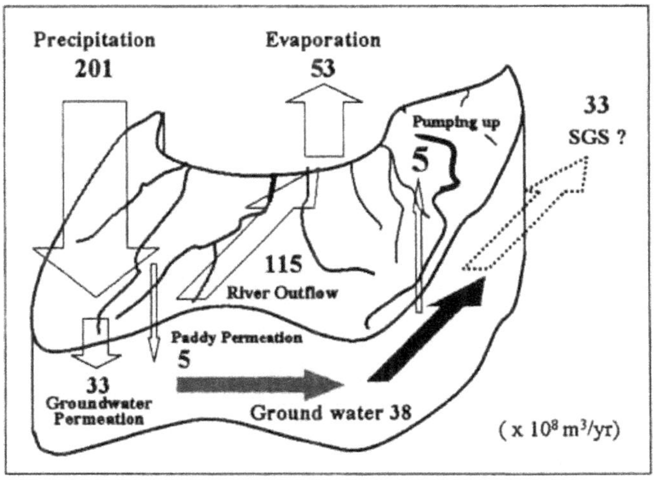

Fig. 8. Water circulation and budget in the Toyama region.

Cable and Taniguchi (2001) indicated that the material flux of the groundwater discharge at the shelf is more important than its contribution to the water balance, because the concentration of dissolved material in the usual groundwater is greater than that of the river water, even if the groundwater runoff to the ocean is merely 10% that of the river. Zektser and Loaciga (1993) reported that the quantities of minerals the groundwater brought to the ocean are half as great as those brought by the river water, even though the groundwater runoff quantity is only 6% that of the river runoff. Though it is said that the Japan Sea Proper Water supplies rich nutrients to Toyama Bay, it is also possible that nutrient-rich groundwater, including the river and the spring at the sea floor, is another source of nutrition. However, the discharge of the groundwater transports to the ocean not only nutrients but also many chemical substances of organic sewage, industrial waste, artificial pesticides draining from paddy fields (Hayashi and Rosenberry, 2001), and so on. Recently, it has been understood that, in the Toyama field, most of the precipitation permeates into the soil, though the concentration of all dissolved carbonates and its $\delta^{13}C$ value come from shallow groundwater (Mizutani and Yamamoto, 1993). Mizutani et al. (2001) indicated the possibility that agricultural water quickly permeates the ground beneath paddy fields in the Kurobe alluvial fan, as there are comparatively small oxygen and hydrogen isotope changes, due to a low evaporation effect. Unlike in river water, where most nutrients and minerals settle to the sediment in the estuary by the salinity gradient, groundwater, with little change in its chemical composition, seems to mix upward into the circumferential seawater because of the water density difference.

The supply of freshwater from a submarine groundwater discharge in the Toyama Bay seems to influence not only the organisms that inhabit the sea floor, but also the circulation of the Japan Sea, as described above. In addition, an investigation of quantitative and qualitative changes of the spring water is in progress, based on detailed information on the geology and hydrology of the land and on the observation area where the spring point extends widely.

Acknowledgments: The authors would like to thank Dr. Taniguchi for coordination and Dr. Gamo for very helpful reviews of this manuscript.

REFERENCES

Back, W. and Hanshaw, B.B., 1970. Comparison of Chemical Hydrogeology of the Carbonate Peninsulas of Florida and Yucatan. Journal of Hydrology 10, 330-368.

Back, W. and Lesser, J.M., 1981. Chemical Constraints of Groundwater Management in the Yucatan Peninsula, Mexico. Journal of Hydrology 51, 119-130.

Barraclough, J.T. and Marsh, O.T., 1962. Aquifers and quality of ground water along the Gulf Coast of Western Florida. 29, State of Florida, State Board of Conservation.

Belanger, T.V. and Walker, R.B., 1990. Ground Water Seepage in the Indian River Lagoon, Florida. In: J.H. Krishna, V. Quinones-Aponte, F. Gomez-Gomez and G.L. Morris (Editors), International Symposium on Tropical Hydrology and Fourth Caribbean Islands Water Resources Congress. American Water Resources Association, San Juan, Puerto Rico, pp. 367-375.

Belanger, T.V., Heck, H.H. and Andrews, M.S., 1997. Groundwater Flow Characteristics of the Mosquito Lagoon, Florida. Project Number: CANA-N-027.000, Florida Institute of Technology.

Bokuniewicz, H., 1980. Groundwater seepage into Great South Bay, New York. Estuarine and Coastal Marine Science 10, 437-444.

Brooks, G.R., Dix, T.L. and Doyle, L.J., 1993. Groundwater/Surface water Interactions in Tampa Bay. Implications for Nutrient Fluxes, The Center for Nearshore Marine Science, University of South Florida.

Burden, D.J., 1964. Karst groundwater investigations, Food and Agric. Org. United Nations (FAO/SF), Greece, pp. 99.

Burnett, W.C., Taniguchi, M. and Oberdorger, J., 2001. Measurement and signification of the direct discharge of ground water into the coastal zone. Journal of Sea Research 46, 109-116.

Cable, J.E., Burnett, W.C. and Chanton, J.P., 1997. Magnitude and Variations of Groundwater Seepage Along a Florida Marine Shoreline. Biogeochemistry 6, 1-17.

Cable, J.E. and Taniguchi, M., 2001. The biogeochemical significance of submarine ground water discharge, Estuaries (for submission).

Cao, H., Cowart, J.B. and Osmond, J.K., 1999. Water Sources of Wakulla Springs, Wakulla County, Florida: Physical and Uranium Isotopic Evidences. Southeastern Geology 39 (1), 51-60.

Cathles, L.M., 1987. Fluid circulation in the crust and the global geochemical budget. In: G.B. Munsch (Editor), Second Conference on Scientific Ocean Drilling (COSOD II). European Science Foundation, Strasborg, France.

Chanton, J.P., Martens, C.S. and Paull, C.K., 1991. Control of pore-water chemistry at the base of the Florida escarpment by processes within the platform. Nature 349 (6309), 229-231.

Church, T.M., 1996. An groundwater route for the water cycle. Nature 380, 579-580.

Cooper, C.R. and Tindall, J.A., 1994. Model for dolomite formation in northwest Florida. Journal of Hydrology 157, 367-391.

Corbett, D.R., 1999. Tracing Groundwater Flow into Surface Waters by Application of Natural and Artificial Tracers. Dissertation for the Ph.D. Degree Thesis, The Florida State University.

Craig, H., 1961. Isotopic Variations in meteoric waters, Science, 133, 1702-1703.

D'Elia, C.F., Webb, K.L. and Porter, J.W., 1981. Nitrate-rich groundwater inputs to Discovery Bay, Jamaica: a significant source of N to local coral reefs? Bulletin of Marine Science 31, 903-910.

Dickson, B.L., 1985. Radium isotopes in saline seepages, south-western Yilgarn, Western Australia. Geochimica et Cosmochimica Acta 49, 361-368.

Dollar, S.J. and Atkinson, M.J., 1992. Effects of Nutrient Subsidies from Groundwater to Nearshore Marine Ecosystems off the Island of Hawaii. Estuarine, Coastal and Shelf Science 35, 409-424.

Fairbridge, R.W., 1966. Submarine Springs IN: The Encyclopedia of Oceanography, Reinhold, New York, 878-883.

Flipse Jr., W.J. and Bonner, F.T., 1985. Nitrogen-Isotope Ratios of Nitrate in Ground Water Under Fertilized Fields, Long Island, New York. Ground Water 23 (1), 59-67.

Fuji, N., 1971. Palynological investigation of the Holocene Uwozu Submerged forest in Toyama Pref.. Central Japan. Bull. Fac. Educat. Kanazawa Univ. 20, 73-87

Fujii, S., 2000. Natural History of Toyama, Memory of the Earth. Katsurashobo, Japan. p. 131,146. (in Japanese).

Fujii, S., Nasu, N., Smith, Alec J., Fuji, N., Mizutani, Y., Shimakura, M., Konishi, K., Igarashi, C., Muramoto, J., Takemura, T., Shimoda, T., Boggs Jr, Sam., Fujioka, K., Mappa, H., Kawahata, H., Kong, Y.S. and Tanaka, T., 1986. Submerged forest off Nyuzen, Kurobegawa alluvial fan, Toyama Bay, central Japan. Boreas l5 (4), 265-277.

Fujii, S., 1992. Toyama Plain. Urban Kubota 4 (31), 40. (in Japanese)

Hanshaw, B.B. and Back, W., 1980. Chemical mass-wasting of the northern Yucatan Peninsula by groundwater dissolution. Geology 8, 222-224.

Hayashi, M., Rosenberry, D.O., 2001. Effects of groundwater exchange on the hydrology and ecology of surface waters. Journal of Groundwater Hydrology. 43 (4), 327-341.

Henderson, G.M., Slowey, N.C. and Haddad, G.A., 1999. Fluid flow through carbonate platforms: constraints from U-234/U238 and Cl in Bahamas pore-waters. Earth and Planetary science Letters 169, 99-111.

Higgins, C.G., 1980. Nips, Notches, and the Solution of Coastal Limestone: An overview of the problem with examples from Greece. Estuarine and Coastal Marine Science 10, 15-30.

Igarashi, G., Saeki, T., Takahata, N., Sano, Y., Sumikawa, K., Tasaka, S., Sasaki, Y., Takahashi, M.,1995. Groundwater radon anomaly before the Kobe earthquake, Science, 269, 60-61.

Ito, T., Fujii, S. 1993. Water budget of groundwater in Toyama basin. Memoirs of the Toyama Geographical Society 10, 3-14.

Johannes, R.E. and Hearn, C.J., 1985. The effect of submarine groundwater discharge on nutrient and salinity regimes in a coastal lagoon off Perth, Western Australia. Estuarine, Coastal and Shelf Science 21, 789-800.

Kanehiro, B.Y. and Peterson, F.L., 1977. Groundwater Recharge and Coastal Discharge for the Northwest Coast of the Island of Hawaii: a Computerized Water Budget Approach. 110, Water Resources Research Center, University of Hawaii, Honolulu.

Komae, T., 1990. Groundwater Research Methods Using Radioactivity. Farming Japan 24(2), 1-7.

Konishi, K., 1983. The Radiocarbon Age of Submerged Forest off Kurobe Alluvial, Toyama Bay. Progress Reports of Low Level Radioactivity Laboratory, Kanazawa University 7, 13. (in Japanese)

Mariotti, A., Landreau, A. and Simon, B., 1988. Nitrogen-15 isotope biogeochemistry and natural denitrification process in groundwater: Application to the chalk aquifer of northern France. Geochimica et Cosmochimica Acta 52, 1869-1878.

Marui, A. and Yasuhara, M., 1999. Studies on Groundwater Flow System with Saltwater-Freshwater Interface. Journal of Japanese Association of Hydrological Sciences 29 (1), 1-12. (in Japanese)

Matson, E.A., 1993. Nutrient flux through soils and aquifers to the coastal zone of Guam (Mariana Islands). Limnology and Oceanography 38 (2), 361-371.

Mizutani, Y. and Satake, H., 1997. Hydrogen and Oxygen Isotope Compositions of River Waters as an Index of the Source of Groundwaters. Journal of Groundwater Hydrology. 39 (4), 287-297. (in Japanese)

Mizutani, Y., Oda, M., 1983. Stable isotope study of ground water recharge and movement in the Shogawa Fan, Toyama. Geochemistry 17 (1), 1-9. (in Japanese)

Mizutani, Y., Satake, H., Yamabe, A., Miyachi, H., Mase, N. and Yamamura, K., 2001. Hydrogen and Oxygen Isotope Ratios of Groundwaters in Shallow Aquifers beneath the Alluvial Fan. Journal of Groundwater Hydrology. 43 (1), 3-11. (in Japanese)

Mizutani, Y., Yamamoto, K., 1993. Sources of dissolved carbonates in the shallow groundwater of the tonami plain, Toyama. Journal of Groundwater Hydrology. 35 (2), 77-86. (in Japanese)

Moore, Y.H., Stoessell, R.K. and Easley, D.H., 1992. Fresh-Water/Sea-Water Relationship Within a Ground-Water Flow System, Northeastern Coast of the Yucatan Peninsula.

Groundwater 30 (3), 343-350.

Muir, K.S., 1986. Groundwater reconnaissance of the Santa Barbara - Montecito Area, Santa Barbara County, California. 1859-A, USGS.

Nakata, T., Tokunaga, T., Asai, K. and Mieta, H., 2001. Nearshore seepage water sampling off Kurobe in Toyama Bay. Chikyu Monthly 23 (12), 857-863.

Nasu, N., Fujii, S. Fujioka, K., Igarashi, C., Kawahata, H., Mappa, H., Konishi, K., Tanaka, T. and Shimoda, T., 1983. Remnants of an ancient forest on the continental shelf of northwest Japan. Boreas 12 (1), 13-16.

Oberdorfer, J.A., Valentino, M.A. and Smith, S.V., 1990. Groundwater contribution to the nutrient budget of Tomales Bay, California. Biogeochemistry 10, 199-216.

Saito, M, Takada, S., Shimamune, T., Nishiki, Y., Shimizu, H. and Hayashi, T., 1996. Tritium electrolytic accumulation using the solid polymer electrolyte. Radioisotopes 45, 285-292.

Sansone, F.J. and Resing, J.A., 1995. Hydrology and geochemistry of sea surface hydrothermal plumes resulting from Hawaiian coastal volcanism. Journal of Geophysical Research 100 (C7), 13555-13569.

Satake, H., 2001. Tritium in precipitation and surface waters. Yoshida, N. (Ed.) Hydrogen and Oxygen Isotopes in Hydrology. Nagoya University and United Nations Educational Scientific and Cultural Organization.

Satake, H., Kanda, Y., 1989. Tritium and Stable Isotope Geochemistry of The Kuranosuke Snow Patch, Central Japan. Annual Report of Tritium Research center, Toyama University, Japan 9, 107-119. (in Japanese)

Satake, H., Kizu, N. and Mizutani, Y., 1984. Tritium Content of River Waters in the Hokuriku District, Japan. Annual Report of Tritium Research center, Toyama University, Japan 4, 67-73. (in Japanese)

Satake, H., Mukai, T. and Mizutani, Y., 1983. Environmental isotope hydrology of precipitations and river waters in the Hokuriku district, Japan. Annual Report of Tritium Research center, Toyama University, Japan 3, 45-56. (in Japanese)

Stoessell, R.K., Ward, W.C., Ford and Schuffert, J.D., 1989. Water chemistry and CaCO3 dissolution in the saline part of an open-flow mixing zone, coastal Yucatan Peninsula, Mexico. Geological Society of America Bulletin 101, 159-169.

Stringfield, V.T. and LeGrand, H.E., 1969. Relation of Sea Water to Fresh Water in Carbonate Rocks in Coastal Areas, with Special Reference to Florida, U.S.A., and Cephalonia (Kephallinia), Greece. J. Hydrol. 9,387-404.

Taniguchi, M. and Fukuo, Y., 1996. An effect of seiche on groundwater seepage rate into Lake Biwa, Japan. Water Resources Research 32 (2), 333-338.

Taniguchi, M. et al., 1999. Disturbances of temperature-depth profiles due to surface climate change and subsurface water flow: 1. An effect of linear increase in surface temperature caused by global warming and urbanization in the Tokyo metropolitan area, Japan. Water Resources Research 35 (5), 1507-1517.

Tokunaga, T., Asai, K., Nakata, T., Taniguchi, M., Shimada, J. and Saegusa, H., 2001. A new technique to collect groundwater samples from submarine formations and its application to offshore Kurobe alluvial fan. Journal of Groundwater Hydrology 43 (4), 279-287. (in Japanese)

Tsunogai, U., Ishibashi, J., Wakita, H., Gamo, T., Masuzawa, T., Nakatsuka, T., Nojiri, Y. and Nakamura, T. 1996. Fresh water seepage and pore water recycling on the seafloor: Sagami

Trough subduction zone, Japan. Earth and Planetary Science Letters 138, 157-158.
Waseda A., Nakai, N., 1983. Isotopic composition of meteoric and surface waters in Central and Northeast Japan. Geochemistry 17 (2), 83-92. (in Japanese)
Whitaker, F.F. and Smart, P.L., 1990. Active circulation of saline ground waters in carbonate platforms: Evidence from the Great Bahama Bank. Geology 18, 200-203.
Wright, D.T., 1999. The role of sulphate-reducing bacteria and cyanobacteria in dolomite formation in distal ephemeral lakes of the Coorong region, South Australia. Sedimentary Geology 126, 147-157.
Zektser, I.S. and Loaiciga, H.A., 1993. Groundwater fluxes in the global hydrologic cycle: past, present and future. Journal of Hydrology 144, 405-427.
Zhang, J., Suzuki, M., Arii, Y., Hattori, H. and Satake, H., 2002. The submarine groundwater seepage off Uwozu in Toyama Bay and their impact on the costal environments. 2002 Japan Earth and Planetary Science Joint Meeting. Tokyo, Japan. H055-008.
Zhang, J., Yamaguchi, Y., Suzuki, Satake, H., Narita, H., Fujita, D., Shozen, K., and Yamaguchi, Y., 2001. Chemical characteristic of the submarine groundwater seepage off Uwozu in Toyama Bay. 2001 Fall Meeting of Oceanographic Society of Japan. Shizuoka, Japan. No. 216. p. 76.

Land and Marine Hydrogeology
Edited by M. Taniguchi, K. Wang and T. Gamo

Prospects of engineering applications of submarine-groundwater-discharge research in Japan

Hideaki Miyamoto and Tomochika Tokunaga

Department of Geosystem Engineering, University of Tokyo, Tokyo 113-8656, Japan

Submarine groundwater discharge (SGD) and chemical fluxes through SGD are subjects of active study in the fields of marine science and hydrology. Although only a few measurements of SGD have been made in Japan, the importance of SGD in coastal hydrological activities has been well recognized in the country. Slight differences in hydrological conditions affecting SGD might sometimes cause serious problems for large-scale constructions such as undersea tunnels and energy storage facilities. Understanding of SGD fluxes and their impact on hydrological environments is important for coastal zone management.

1. INTRODUCTION

Mountainous regions cover about 80% of the total land area of Japan, and a large portion of the Japanese population of 125 million live in narrow plain areas close to shorelines. The high population density has been causing engineering and environmental problems in the coastal areas, such as refuse disposals, traffic congestions, and disruption of natural environments. Many reclaimed lands have been formed in the Tokyo Bay area. Some of the reclaimed lands have been filled with wastes. Additional engineering problems have been caused by recent construction of public facilities on the reclaimed lands (Shimizu et al., 1992). Recent and rapid increases in utilization of the reclaimed areas have brought about strong needs for a comprehensive understanding of hydrological conditions and their temporal changes (see Endo et al. (2001) for a recent review of the historical changes of the groundwater condition in the Tokyo area). To evaluate the water condition in a coastal region, we should make good estimates of riverine water fluxes, meteologic inputs, groundwater movements, and distribution of geological formations.

Recently submarine groundwater discharge (SGD) has been studied actively for its potential as a significant source of water/chemical discharge into the sea. In some cases, the total SGD fluxes are reported to be several tens of percent of the total flow entering the sea from adjacent rivers (e.g., Moore, 1996). The influence of SGD on the global hydrologic condition is now actively discussed and will be a major subject of study in submarine and coastal hydrology. Distribution of saltwater and freshwater, which should be related to SGD, in the coastal zone has also been studied from an engineering point of view. In this paper, we briefly review recent studies on SGD and its role in hydrological environments. We also review some problems and interesting topics of SGD research in Japan from an engineering point of view. Tectonic setting of the Japanese Island and localities of undersea tunnels, undersea coal mines, energy storage caverns, and active volcanoes discussed in this article are shown in Figure 1.

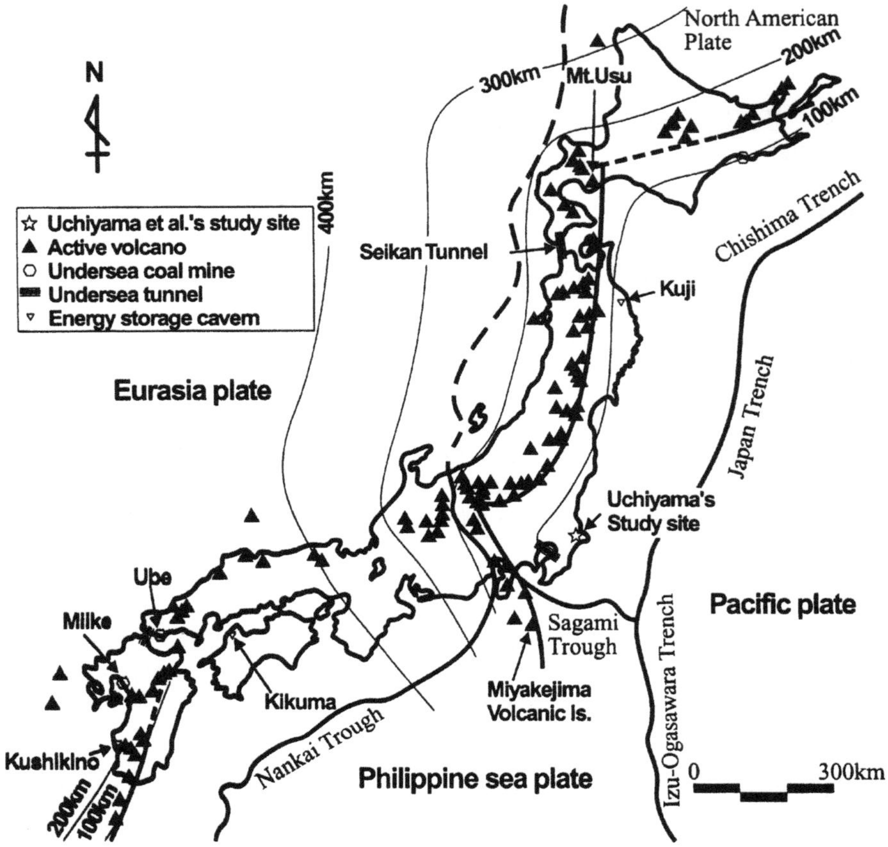

Figure 1. Tectonic setting of the Japanese islands and localities of undersea tunnels, undersea coal mines, energy storage caverns and active volcanoes discussed in the text. Modified after Committee for Catalog of Quaternary Volcanoes in Japan (1999). Contour lines indicate the inferred depth of the subducting slab.

2. SGD AND ITS ROLE IN HYDROLOGICAL ENVIRONMENTS

In contrast to major rivers that are gauged, local or global SGD into the ocean remains difficult to evaluate. This is due mainly to the difficulty in measuring SGD. Some workers have carefully estimated SGD fluxes in some areas and concluded that the global SGD flux might be considerably lower than the riverine flux (e.g., Zektser and Loaiciga, 1993; Cable et al., 1996a). However, studies in some other areas have indicated that SGD may occasionally account for a significant fraction of the fresh water inflow (e.g., Valiela and D'Elia, 1990; Moore, 1996; Church, 1996). These results were thought to suggest that the local SGD fluxes in different coasts were considerably different.

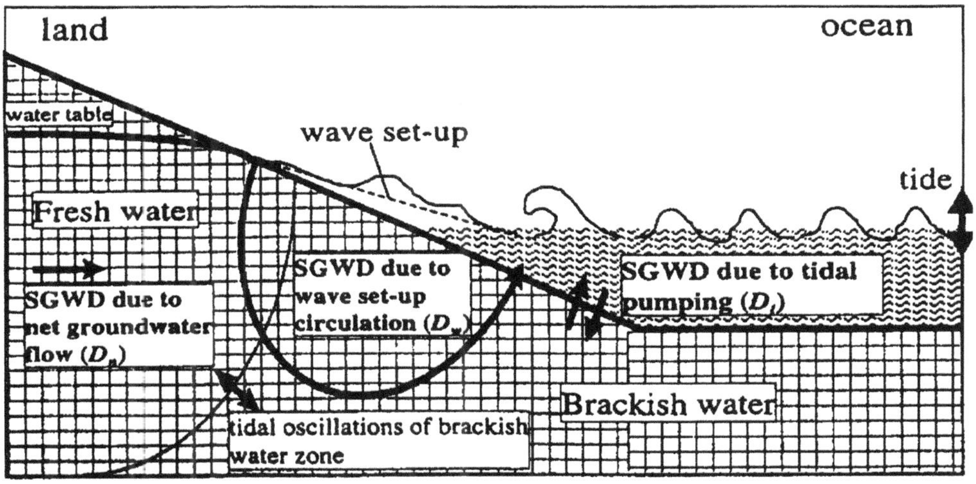

Figure 2. Schematic illustration of processes affecting submarine groundwater discharge. After Li et al. (1999).

However, recent understanding of the hydrological processes in coastal zones has made it possible to explain the discrepancy in a different way. In this paper, we follow the discussion by Li et al. (1999), who used the definition of SGD by Church (1996). According to Church (1996), SGD is the direct groundwater outflow across the ocean-land interface into the ocean. In the coastal zone, there exist several processes which affect SGD (Figure 2). Not only the net fresh groundwater discharge (D_n), but also groundwater circulations driven by the potential gradient due to wave setup (D_w) and tidal pumping (D_t) contribute to SGD. Li et al. (1999) proposed that the amount of total discharge (D_{SGD}) could be written as:

$$D_{SGD} = D_n + D_w + D_t \, . \tag{1}$$

D_w and D_t are considered to be primarily of marine origin. The recent field experiment by Moore (1996) indicated that D_{SGD} was about 40 % of riverine discharge at the studied coast (South Atlantic Bight). Younger (1996) argued that only 4 % of the D_{SGD} was the net fresh groundwater discharge, suggesting that a very large portion (about 96 %) of the measured D_{SGD} was recirculated at the land-ocean interface. The theoretical model for the SGD and associated chemical transfer to the ocean by Li et al. (1999) supports the field measurements by Moore (1996) and the argument by Younger (1996). This understanding is important in evaluating the measurements and estimations of SGD.

2.1. Measurements and estimations of SGD

There are basically two approaches for SGD assessments (e.g., Moore, 1999):

1. **Direct measurement:** Direct measurements of SGD are made at the sea floor using seepage flux meters or by determining the direction and magnitude of the hydraulic gradient across the sediment-water interface (e.g., Taniguchi and Fukuo, 1993). The in-situ approaches provide precise flux. However, they give only very localized flux estimates and require further improvements for a regional flux evaluation.
2. **Indirect method**: Indirect measurements of groundwater flow into the coastal sea use geochemical species which are enriched in groundwater and behave conservatively in seawater (e.g., Moore, 1996; Cable et al., 1996b; Shimada et al., 1993). Artificial tracers are also measured after their injection into groundwater systems for the evaluation of flow rates and directions. Well tests and other hydrological measurements have also been attempted to

give ground-truth information for several modeling approaches ranging from simple on-shore groundwater balance calculations to complex numerical models of sub-surface flow.

Both of the above methods are widely used. It should be noted that we must understand exactly what we measure (or estimate) using the above mentioned methods. For example, the seepage flux meters measure D_{SGD} which include both net fresh groundwater discharge and recirculated seawater. Thus, we need other information to differentiate net groundwater discharge from recirculated seawater (Moore, 1999). Geochemical methods also need to be carefully evaluated because a small change in groundwater condition changes the behavior of some chemical species. It is known that Radium, Cesium, phosphate, ammonia, and cadmium tend to be absorbed by sand particles in fresh groundwater and to desorb into brackish groundwater. Thus, a small rate of seawater intrusion will result in significant chemical transfer to the ocean (Li et al., 1999). As a temporal process due primarily to wave setup and tidal pumping, it might not be relevant to significant discharge of fresh groundwater. On the other hand, D_{SGD} estimates using water balance and numerical modeling of groundwater flow evaluate only the net fresh groundwater discharge to the ocean. Thus, understanding the characteristics and the strengths and weaknesses of different approaches is necessary. In this sense, intercalibration experiments suggested by Burnett et al. (2001) are an important next step.

In Japan, there have been very few studies related to SGD. Uchiyama et al. (2000) studied the nutrient flux at the coastal zone through field measurements and numerical simulation and showed that the nutrient flux via groundwater seepage was responsible for a relatively minor component of productivity in the surf zone. Their results imply that the fraction of SGD in total water/chemical flux into the sea might be relatively minor in Japan. Their study area is, however, a zone in which a big river runs sub-parallel to the coast, forming a narrow isolated hydrological system (the width is less than 10km). Thus, their result may not represent the average nutrient flux by SGD in Japan. In any case, we need to understand the importance of SGD in local hydrological activities in individual coastal areas to be able to manage coastal aquifers, to prevent seawater intrusion due to anthropogenic activities, and to evaluate groundwater pollution (especially when SGD acts as a pathway for contaminant fluxes into the sea).

2.2. Shape of saltwater-freshwater interface

Although it may not be directly related to SGD, the shape of saltwater-freshwater interface and its temporal change have been key factors in several engineering activities in Japan, and will be related to the evaluation of future projects such as those for nuclear waste disposal. Seawater intrusion is the migration of seawater into freshwater aquifers that are in hydraulic connection with the sea and under the influence of groundwater extraction. Intrusion of seawater into coastal aquifers is an important process that leads to groundwater salinization to levels exceeding those acceptable for drinking and irrigation. This problem was extensively studied in the past. Original works on the shape of freshwater-saltwater boundary include those by Badon-Ghyben (1888) and Herzberg (1901). See Bear et al. (1999) for a comprehensive treatment of seawater intrusion problems.

The change of Earth's climate seems to affect strongly the shape of freshwater-saltwater boundary. The shoreline of the continental shelf has retreated more than several tens of kilometers from the last glacial period to the present, due to sealevel rise of more than 100 m. This large amount of sealevel rise has a serious impact on the boundary between the freshwater and saltwater. The speed of sealevel rise in a postglacial period is much faster than the speed of sealevel subsidence during a glacial period. Therefore, some freshwater might remain below the present seafloor. Deposition of clayey sediments during the sealevel rise may impede the rapid infiltration of seawater. Kooi et al. (2000) conducted a numerical sensitivity study on this problem and found that, depending on the values of hydraulic conductivity and aquifer/aquitard distribution, significant amount of freshwater/brackishwater could exist beneath the continental shelves and shallow seas. Especially, if low-permeablity sediments exist at the seafloor, freshwater tends to remain in subsea formations. Studies on water chemistry offshore of New Jersey (Hathaway et al., 1979) and Suriname (Groen et al., 2000) clearly showed the existence of a broad zone of fresh to brackish pore water under the continental shelf (Figures 3 and 4), pointing to the importance of long-term geological processes on the hydrology of coastal areas and continental shelves. Hathaway et al. (1979) also noted the existence of a Quaternary aquitard in the top sedimentary unit and emphasized the importance of this aquitard for the observed freshwater zone at the New Jersey shelf. The freshwater pockets observed during previous engineering activities in Japan (see also sections 3.1 and 3.2 for Seikan Tunnel and underground coal mining) may also be remnant groundwater from the last glacial period. One point that should be considered for groundwater displacement on this timescale is that the amount of rainfall changes according to the change of surface climate condition. Tsukada (1983) studied pollen assemblages in Japan and concluded that the probable annual precipitation ranged from 1050 to 1300 mm along Pacific coasts in southwestern Japan during the last glacial maximum. This precipitation is about 40 to 70% of the present annual precipitation. Although recharge area for groundwater increases in a glacial period due to ocean regression, the amount of recharge per unit area must decrease due to the smaller amount of rainfall. Combined effects of changing surface environments during glacial/interglacial periods due to changes in shoreline, precipitation, sedimentation/erosion, etc., on the groundwater budget are not yet clearly understood.

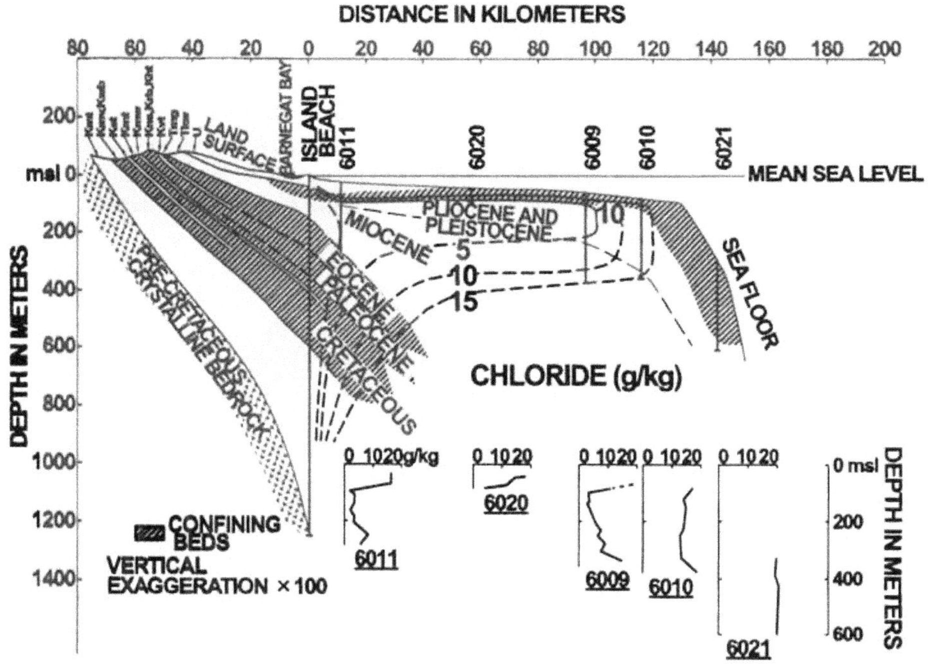

Figure 3. Hydrological cross section of the coastal plain of New Jersey and the continental shelf. Dashed contours show the chlorinity of pore water. After Hathaway et al. (1979).

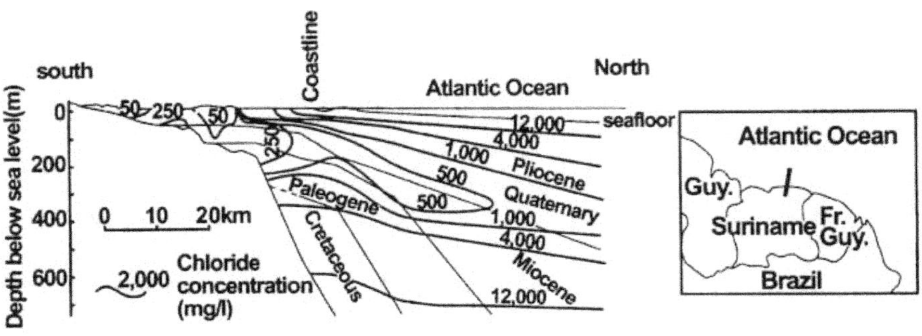

Figure 4. Simplified hydrological cross section of the coastal plain of Suriname and the continental shelf. After Kooi et al. (2000).

3. SGD AND LARGE SCALE CONSTRUCTIONS

As discussed in the previous section, SGD is directly related to the local hydrological conditions. Small differences in the conditions might sometimes cause a serious problem to construction activities. In this section, we focus on some reported problems in Japan that are actively discussed in engineering hydrology.

3.1. Undersea tunnels

During the 20th Century, several undersea tunnels were constructed in Japan. Constructions and maintenances of undersea tunnels strongly demand comprehensive understandings of hydrologeologic environments. The Seikan Tunnel, the longest undersea tunnel in Japan, connects the Honshu island to the Hokkaido Island (Figure 5). The overall length of the tunnel is 53.85 km, of which 23.35 km is situated beneath the sea floor. Construction of the Seikan Tunnel was a great challenge because of serious geological and hydrological conditions. In particular, catastrophic discharge of water into the tunnel was one of the most serious problems. Severl to 80 tons per minute of water discharge were observed in several places during the excavation of pilot tunnels, which significantly delayed the progress of tunnel construction (Seki, 1980).

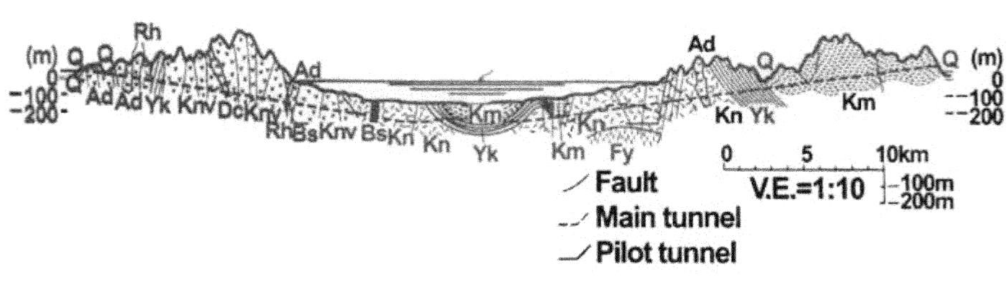

Figure 5: Simplified geological cross section of the Seikan Tunnel. Abbreviations are, Q: Quaternary formations, Km: Kuromatsunai Formation, Yk: Yakumo Formation, Kn: Kunnui Formation, Knv: Kunnui Formation (volcaniclastic sediments rich part), Fy: Fukuyama Formation, Rh: Rhyolite, Dc: Dacite, Ad: Andesite, Bs: Basalt. After Sasa (1972).

The construction of an undersea tunnel below the Tsugaru Strait was carried out through incipiently metamorphosed Miocene volcaniclastic sediments. The water depth of the Tsugaru Strait is less than 140 m, and the area was exposed to surface during the last glacial maximum (Sasa, 1972). The formation water (pore fluid) is fresh and is considered to be the fresh water from the last glacial maximum. Salt water from sea bottom has been displacing the formation water through fractures. The alternative hypothesis that the present freshwater hydrologic system may extends to below the sea bottom is considered improbable due to the lack of a strong driving mechanism.

Seki (1980; 1981) analyzed the water-rock interaction and the resulting water chemistry from the Seikan Tunnel and proposed a technique to predict disastrous entry of seawater using water chemistry. According to his study, the volcaniclastic rocks have experienced diagenesis and hydrothermal alteration. The main altered mineral is montmorillonite, rich in calcium and magnesium but poor in sodium and potassium. The cation exchange between the groundwater and this clay mineral is an important chemical process that alters the water chemistry. This reaction can be written as:

$$\text{Ca-montmorillonite} + (Na^+, K^+) \rightarrow \text{Na K-montmorillonite} + Ca^{2+} \quad (2)$$

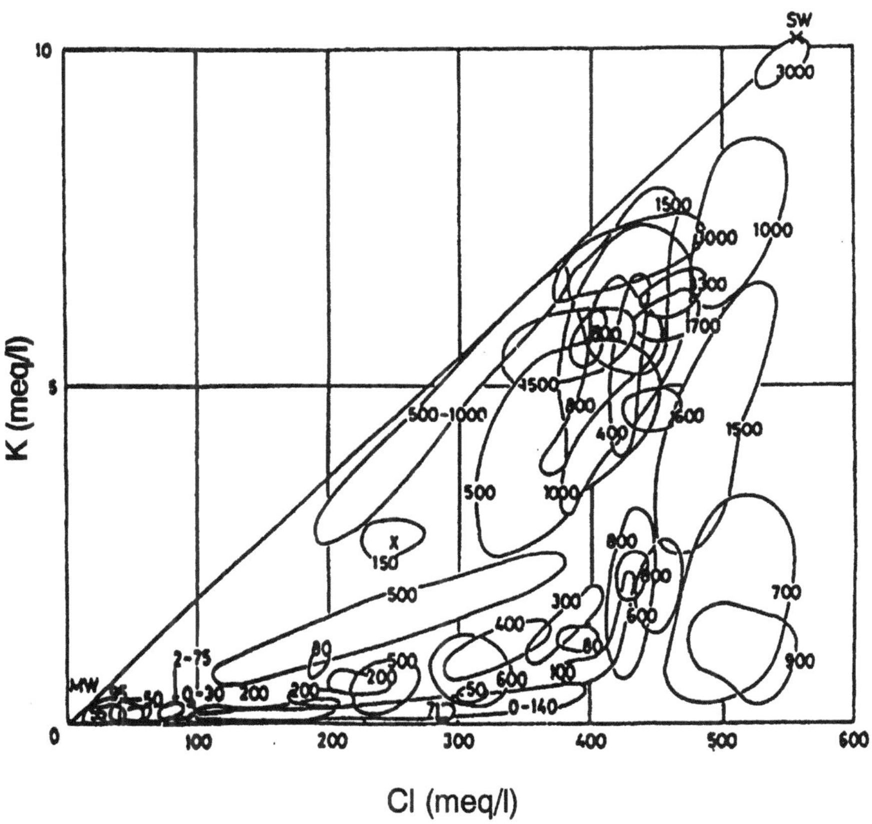

Figure 6. Observed water discharge rates (l/min) during the construction of undersea part of the Seikan Tunnel on the K-Cl diagram. SW: seawater, MW: meteoric water. See text for detailed discussion. After Seki (1981).

Thus, the potassium and sodium content decreases with increasing degree of water-rock interaction. Considering that the potassium content is much smaller than the sodium content in sea water, the potassium content can be a good indicator of the degree of cation exchange. On the other hand, the chlorite content does not change due to water-rock interaction, and it indicates the mixing ratio between the seawater and fresh pore water. Seki (1981) introduced the K-Cl diagram to indicate the potential of rapid discharge of sea water into the tunnel. The idea behind it is that the increase in the chloride content suggests an increase in sea water intrusion, and the decrease in the potassium content suggests an increase in water-rock reaction time. Thus, an increase in both chloride AND potassium is the best warning signature. A chloride increase without a simultaneous potassium increase indicates that the discharged water is of sea water origin but has passed through rock formations over a long time (through porous matrices), suggesting that the intake of sea water through connected fractures does not take place. Figure 6 is the comparison of measured discharge rate with chloride-potassium concentration, and the comparison is consistent with Seki's idea.

This technique can be very helpful for construction projects like the Seikan Tunnel. However, water-rock interaction depends strongly on rock types and reaction sequence (the order in which various minerals or groups of minerals are encountered by the water that moves through the flow system; see Freeze and Cherry (1979) for details), and hence the technique cannot be generalized as an effective prediction strategy for undersea tunnels. It is necessary to understand the details of the flow system of the area of interest to establish site-specific prediction methods.

3.2. Undersea coal mining

Hydrogeological conditions discussed in the preceding section have also been encountered in undersea coal mines in Japan. Main coal producing fields in Japan are situated in the coastal region and subsea formations (See Figure 1 for the locations of major coal fields). These coal mines extend about 10 km seaward of the coastal line, under water depths of several tens of meters. Water is discharged into underground coal fields, and seepage is one of the largest factors contributing to mining tunnel collapse. The chemistry of the discharged water has been used to assess the safety of coal mining activities. Because coal mining activities generally continue for more than several tens of years, relatively large amounts of data on the distribution of water chemistry have been gathered by engineers of coal mines. According to these data, formation water of Paleogene formations (including workable coal seams) tends to have a lower chloride concentration than sea water. This situation might be due to the existence of impermeable fault zones and overlying clayrich Quaternary sediments of several hundred meters thick (Iwasawa and Ueda, 1961; Kikuchi et al., 1971). At the Miike coal field, the coal bearing formation is seaward dipping, and mining tunnels also become deeper in the seaward direction. Figure 7 shows that the chloride concentration drops at a depth of 180 m where the tunnels cross one major fault, suggesting that the fault acts as an effective barrier for the migration of formation water (Kikuchi et al., 1971). Formation water beyond the fault is little affected by the intrusion of seawater. A very similar areal distribution of chloride concentration was reported by Tanaka (1999) using data from Ube coal field.

The above mentioned situation makes it possible for engineers of coalmines to use water chemistry, mainly chloride concentration, as a measure of the possibility of seawater intrusion (Iwasawa and Ueda, 1961).

3.3. Storage of energy

For an effective utilization of the surplus power at nighttime and during holidays, the feasibility of power storing facilities is actively investigated. Among them, CAES-G／T （Compressed Air Energy Storage-Gas Turbine） power generation is one of the most attractive solutions because of its efficiency in storage and maintenance. The storage facility is planned to be installed in bedrock formations close to urban areas, and detailed and precise understanding of the hydrological environment is strongly demanded.

In addition, storage facilities of emergency energy and a system to distribute energy supplies efficiently are also important in Japan, because in Japan almost all fuels, such as petroleum, liquefied petroleum gas (LPG), and liquefied natural gas (LPG), are imported. Three underground oil stockpiling plants have been constructed in coastal areas in Japan. Construction and maintenance of huge storage tanks in underground rock formations such as Kuji (1.75Mkl for storage capacity), Kikuma (1.50Mkl), and Kushikino (1.75Mkl), also strongly require a comprehensive understanding of the hydrological environment. These facilities are developed several tens of meters below the ground surface near the shoreline, and therefore groundwater salinity is one of the primary concerns. A water sealing technique is often used for these facilities. Water sealing may significantly influence local groundwater flow,

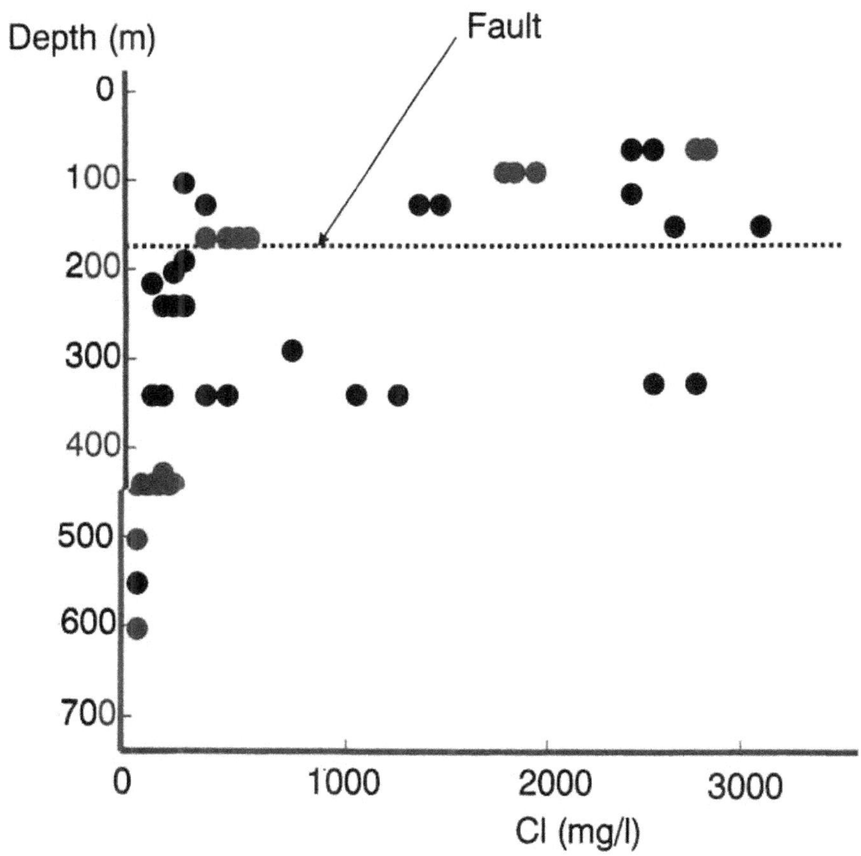

Figure 7. The relationship between the depth below sea surface and chloride concentration at the Miike coal field. Modified after Kikuchi et al. (1971).

which may be observed as changes in SGD fluxes or locations. For detection of possible leakage, SGD fluxes and their chemistry can be monitored.

3.4. Radioactive waste-disposal facility

For planning the possible construction of radioactive waste-disposal facilities, long-term stability and possible evolution of the hydrological environment have been considered. These facilities have to isolate radioactive nuclides from surface biosphere for a rather long period.

Near-coast sub-sea area was first mentioned by Chapman et al. (1986) as one of the suitable environments for disposal of long lived wastes in the United Kingdom. The main reason is that groundwater movements are expected to be very slow under the coast because of the perceived lack of head variations in subsea formations. If the disposal facility is constructed seaward of the freshwater/saltwater interface, the migration of nuclides is expected to be diffusion dominated, because topographic drive of groundwater flow is small.

Japan Nuclear Cycle Development Institute (hereafter called JNC) conducted a generic study on the effect of the change of pore fluid condition (saltwater to freshwater due to possible regression) on the overall performance of a disposal system that includes both an engineering barrier system and far-field geological formations as long-term barriers for the migration of radionuclides (JNC, 1999). Sealevel changes affect the hydrological environment, especially the distribution of fresh and salty pore fluids, in the coastal areas and continental shelves. The change of pore fluid composition from saltwater to freshwater affects the solubility and distribution coefficients of radionuclides and hence the migration efficiency. It should be noted that some radionuclides move more rapidly (e.g., ^{237}Np) due to these changes, but others tend to move more slowly because of the decrease in solubility (e.g., ^{79}Se) and the increase in distribution coefficient (e.g., ^{135}Ce) (JNC, 1999). JNC (1999) concluded that the change of pore fluid conditions would not strongly affect the maximum calculated dose at the bioshpere, and the calculated value was very similar to that obtained using saltwater conditions throughout the evaluated time period. Hasegawa et al. (2001) recently conducted two-dimensional numerical sensitivity studies to evaluate the effects of sea level changes on the distribution of different pore fluids (freshwater and saltwater) and on the temporal change of fluid flow velocity. Their results suggest that the flow velocity is generally slower in a saltwater region than in a freshwater region by one or two orders of magnitude. Their results also predict significant movement of freshwater/saltwater boundary at a continental shelf setting due to sea-level change. Their numerical calculation, however, does not include the change of surface conditions such as deposition and/or erosion of geological formations. Considering that the existence of thin low permeability sediments strongly affects the displacement pattern of freshwater/saltwater in pore space (Kooi et al., 2000), it is desirable to conduct modeling that includes the above mentioned geological processes. In this sense, studies on submarine groundwater discharge, long-term change of hydrological conditions in the costal area, and related hydrological/geological phenomena are important from the engineering point of view.

4. NATURAL HAZARDS PREVENTIONS

SGD is possibly correlated to some natural hazards such as volcanic eruptions, subaqueous landslides, and earthquakes. Weak correlations are suggested by some workers. Studies of these correlations may help understand or predict natural hazards and their effects on hydrological systems.

4.1. Volcanic hazard

Some recent investigations revealed interactions between groundwater and active magmas.

For example, a remarkable increase in groundwater temperature was observed on Izu-Oshima island about half a year after the 1986 eruption. A Submarine eruption several kilometers away from the east coast of the Izu Peninsula occurred in 1989. Geochemical and hydrological observations at hot spring wells in Ito City indicated that water level, temperature and chemical composition of a hot spring had changed during the 1989 volcanic event (Notsu et al., 1991; Sato et al., 1992). Most of the eruptions at Mt. Usu in 2000 were phreatic explosions, blasts generated by steam produced from magma-heated groundwater under and around the volcano. Discolored seawater was found on the western submarine-flank about 1.8 kilometers from the coast of Miyakejima volcanic island, suggesting a submarine eruption associated with a volcanic eruption in 2000. A remotely controlled camera was used to try to find SGD, but it could not detect any SGD in this area. However, this type of research might contribute to future predictions of eruptions.

Although effects of groundwater on volcanic activities are believed to be significant, especially for volcanic islands, hydrothermal and groundwater conditions at and near volcanic islands are poorly understood. Continuous observations of SGD at volcanic islands will help understand these conditions.

4.2. Subaqueous landslide and tsunami

Subaqueous landslides are a potential hazard for telephone lines, undersea constructions, fish and pearl farms, and other submarine facilities. There are a number of factors that could trigger a slope failure, such as the desociation of gas hydrates due to warmer ocean temperatures, but SGD is suggested to be one of the most possible trigger (e.g., Anthony and Julian, 1997). Sudden groundwater discharge is particularly dangerous because it can cause big subaqueous landslides. Potential submarine landslides on the outer continental shelf could trigger tsunamis that might have devastating effects on populated coastal areas (Ward, 2001). Driscoll et al. (2000) newly discovered cracks along the edge of the continental shelf that could be an early warning sign that the seafloor is unstable in these areas. Similar signs may be found in other coastal areas.

5. OTHER ASPECTS

5.1. Pollution

Economic development has been most active in coastal zones, putting enormous pressure on coastal ecosystems. In the 1960s, heavy industries concentrated along Japanese coastal areas caused extreme water pollution. This was serious especially in semi-closed bay areas and led to damages to fishery resources. Although strict enforcement of laws and standards since the 1970s has improved the quality of coastal waters, eutrophication in these areas is still a serious problem.

Coastal and marine water pollution has increased throughout the region, due mainly to direct discharge from polluted rivers, increased surface run-off and drainage from expanding port areas, oil spills and other contaminants from shipping and industries. The flux through SGD should be precisely measured because of its potential of being a contaminant pathway.

5.2. Possibility of new freshwater resources

Lack of drinking and industrial water is a serious problem especially in volcanic islands in Japan, because these islands depend nearly solely on groundwater as a source of fresh water. Growing economy on thse islands needs much more fresh water. However, excessive use of groundwater will cause seawater intrusion into aquifers and sometimes land subsidence in a coastal area. Because the quality of SGD is similar to that of pure groundwater, SGD may be used instead of deep groundwater as a new water resource (Marui, 1997). To pursue this interesting possibility, further research is needed especially for understanding the locations and fluxes of SGD and their flow systems.

6. CONCLUSION

The significance of SGD is now being recognized in engineering fields in Japan. There are many reported problems and interesting topics concerning SGD in engineering fields. However, basic facts such as SGD locations and their fluxes are still poorly quantified. Further scientific and engineering investigations will allow these issues to be addressed with a higher degree of certainty. Especially, local and detailed monitoring is quite important and should be enhanced.

Acknowledgements. We would like to thank Alexis Palmero and Taku Iwai for their informal reviews and helpful comments on an earlier version of the manuscript. This manuscript benefited greatly from the careful and constructive reviews by editor Kelin Wang and two anonymous reviewers.

REFERENCES

Anthony, E.J., Julian, M., 1997. The 1979 Var Delta landslide on the French Riviera: A retrospective analysis. J. Coastal Res. 13, 27-35.

Badon-Ghyben, W., 1888. Nota in verband met de voorgenomen putboring nabij Amsterdam. Tijdschr. Inst. Ing. 9, 8-22.

Bear, J., Cheng, A.H., Sporek, S., Ouazar, D., Herrera, I., 1999. Seawater intrusion in coastal aquifers - concepts, methods, and practices. In: Dordrecht. Kluwer Academic Publishers. 625pp.

Burnett, W.C., Taniguchi, M., Oberdorfer, J., 2001. Measurement and significance of the direct discharge of groundwater into the coastal zone. J. Sea Res. 46, 109-116.

Cable, J.E., Bugna, G.C., Burnett, W.C., Chanton, J.P., 1996a. Application of ^{222}Rn and CH_4 for assessment of groundwater discharge to the coastal ocean. Limnol. Oceanogr. 41, 1347-1353.

Cable, J.E., Burnett, W.C., Chanton, J.P., Weatherly, G.L., 1996b. Estimating groundwater discharge into the northeastern Gulf of Mexico using radon-222. Earth Planet. Sci. Lett. 144, 591-604.

Chapman, N. A., McEwen, T.J., Beale, H., 1986. Geological environments for deep disposal of intermediate level wastes in the United Kingdom. In: Proceedings of the International Symposium on the Siting, Design and Construction of Underground Repositories for Radioactive Wastes, IAEA, Hanover, 3-7, March 1986. 311-328.

Church, T. M., 1996. An underground route for the water cycle. Nature. 380, 579-580.

Committee for Catalog of Quaternary Volcanoes in Japan, 1999. In: The Volcanological Society of Japan (Ed.) Catalogue of Quaternary Volcanoes in Japan. CD-ROM (in Japanese).

Driscoll, N.W., Weissel, J.K., Goff, J.A., 2000. Potential for large-scale submarine slope failure and tsunami generation along the US mid-Atlantic coast. Geology. 28, 407-410.

Endo, T., Kawashima, S., Kawai, M., 2001. Historical review of development of land subsidence and its cease in Shitamachi Lowland, Tokyo. J. Jpn. Soc. Eng. Geol. 42, 74-87 (in Japanese with English abstract).

Freeze, R. A., Cherry, J.A., 1979. In: Groundwater. Prentice-Hall, Englewood Cliffs, N. J. 384-457.

Groen, J., Velstra, J., Meesters, A.G.C.A., 2000. Salinization processes in paleowaters in coastal sediments of Suriname: evidence from δ 37Cl analysis and diffusion modeling. J. Hydrol. 234, 1-20.

Hasegawa, T., Igarashi, T., Tanaka, Y., Tsuchi, H., Ono, F., 2001. Numerical investigation of sub-seabed groundwater flow. J. Ground. Hydrol. 43, 289-300 (in Japanese).

Hathaway, J.C., Poag, C.W., Valentine, P.C., Miller, R.E., Shultz, D.M., Manheim, F.T., Kohout, F.A., Bothner, M.H., Sangrey, D.A., 1979. U. S. Geological Survey core drilling on the Atlantic shelf. Science. 206, 515-527.

Herzberg, A., 1901. Die Wasserversorgung einiger Nordseebader. J. Gasbeleucht. Wasserversorg. 44, 815-819.

Iwasawa, S. Ueda, H., 1961. Seepage in the minning tunnel of Ube-mining industry. J. Jpn. Min. Indust. 77, 341-348.

Japan Nuclear Cycle Development Institute (JNC), 1999. H12 : Project to Establish the Scientific and Technical Basis for HLW Disposal in Japan. Japan Nuclear Cycle Development Institute.

Kikuchi, H., Kondo, T., Kuroda, K., 1971. Groundwater in the Miike coal mining tunnel. In: Matsumoto, T. (ed.), Professor Hisamichi Matsushita Memorial Volume. II, Fukuoka Insatsu, Fukuoka, 217-231 (in Japanese with English abstract).

Kooi, H., Groen, J., Leijnse, A., 2000. Modes of seawater intrusion during transgressions. Water Resour. Res. 36, 3581-3589.

Li, L., Barry, D.A., Stagnitti, F., Parlange, J.Y., 1999. Submarine groundwater discharge and associated chemical input to a coastal sea. Water Resour. Res. 35, 3253-3259.

Marui, A., 1997. Submarine groundwater discharge -possibility of opening up new resources. J. Hydrol. Sci. 27, 85-94.

Moore, W., 1996. Large groundwater inputs to coastal waters revealsed by ^{226}Ra enrichments. Nature. 380, 612-614.

Moore, W.S., 1999. The subterranean estuary: A reaction zone of ground water and sea water. Marine Chem. 65, 111-125.

Notsu, K., Wakita, H., Igarashi, G., Sato, T., 1991. Hydrological and geochemical changes related to the 1989 seismic and volcanic activities off the Izu Peninsula. J. Phys. Earth. 39, 245-254.

Sato, T., Wakita, H., Notsu, K., Igarashi, G., 1992. Anomalous hot-spring water changes - possible precursors of the 1989 volcanic eruption off the east coast of the Izu Peninsula. Geochem. J. 26, 73-83.

Sasa, Y., 1972. Marine geology of the western part of the Tsugaru Strait. Kagaku. 42, 40-50 (in Japanese).

Seki, Y., 1980. Geochemistry of water rock interaction and prediction of gushing water in undersea tunnel construction. Ann. Rep. of Exp. Facil. of Rock and Hydrol. Facul. Eng., Univ. Saitama. 6, 34-39 (in Japanese).

Seki, Y., 1981. An example of the prediction of the seawater gushing into undersea tunnel by chemical analysis of seepage. Ann. Rep. of Exp. Facil. of Rock and Hydrol. Facul. Eng., Univ. Saitama. 7, 6-18.

Shimada, J., Kayane, I., Shimano, Y., Taniguchi, M., 1993. Use of several environmental tracers to detect the surface-subsurface water interaction in an alluvial fan. Tracers in Hydrol. 215, 263-274.

Shimizu, K., Tokunaga, T., Kojima, K., 1992. Metropolitan urban geology. 29th Intern. Geol. Cong. Field Trip Guide A13. 13pp.

Tanaka, K, 1999. Groundwater flow at the coastal areas. Proceedings of 1999 Annual Symposium of the Japan Society of Engineering Geology, Japan Society of Engineering Geology, Tokyo. 29-38 (in Japanese).

Taniguchi, M., Fukuo, Y., 1993. Continuous measurements of grouns-water seepage using an automatic seepage meter. Ground Water. 31, 675-679.

Tsukada, M., 1983. Vegetation and climate during the Last Glacial Maximum in Japan. Quarter. Res. 19, 212-235.

Uchiyama, Y., Nadaoka, K., Rolke, P., Adachi, K., Yagi, H., 2000. Submarine grondwater discharge into the sea and associated nutrient transport in a sandy beach. Water Resour. Res. 36, 1467-1479.

Valiela, I. D'Elia, C., 1990. Groundwater inputs to coastal waters – introduction. Special Issue Biogeochem. 10, 175-175.

Ward, S.N., 2001. Landslide tsunami. J. Geophys. Res. 106, 11,201-11,215.

Younger, P.L., 1996. Submarine groundwater discharge. Nature. 382, 121-122.

Zektser, I.S., Loaiciga, H., 1993. Groundwater fluxes in the global hydrological cycle: Past, present and future. J. Hydrol. 144, 405-427.

Land and Marine Hydrogeology
Edited by M. Taniguchi, K. Wang and T. Gamo

Evaluation of seawater intrusion accompanying the coastal coalmine excavation in the Joban coalfield area, Japan

J. Shimada [a], K. Kojima [b], K. Ohara [c] and M. Yamakawa [d]

[a] Dept. of Earth Science, Kumamoto University,
2-39-1 Kurokami, Kumamoto, 860-8555, Japan

[b] Emeritus Professor, University of Tokyo, & President of Geospace Labo.,
4-2-6 Ginza, Chuo-ku, Tokyo 104-0061, Japan

[c] President, Chishitsukiso-kogyo Co.,
3-163-1 Uchiko Miumaya-cho, Iwaki, 973-8403, Japan

[d] Japan Marine Science and Technology Center
2-15 Natsushima-cho, Yokosuka, 237-0061, Japan

The change of groundwater conditions during the time of coastal coal mining excavations (down to the depth of 1000m) and after their abandonment has been studied in the Tertiary sedimentary formations with low permeability and little groundwater data existing. The actual groundwater disturbance caused by the mine excavation in the coastal sedimentary bedrock has been studied based on the information during the operational periods of the Joban coalmine over about the past 100 years, and the recently sampled data from the subsequent hot springs in the study area.

The study provides important information about seawater intrusion caused by the groundwater disturbance from long-term and large-scale underground excavations.

1. INTRODUCTION

The fresh-salt water interface zone in the coastal aquifer exists, and many submarine groundwater springs are reported at their mixing zone (Marui and Yasuhara, 1999, and Hayashi *et al.*, 1999) because of the density difference between seawater and fresh groundwater. These submarine groundwater springs are mainly observed in the hydrologically active coastal groundwater aquifer especially in the unconsolidated sand and gravel aquifers. Little information on these interface zones and the submarine groundwater springs has been known in the consolidated sedimentary bedrock aquifers with low permeability. Even the groundwater flow system in this sedimentary bedrock aquifer has not been clarified.

Large-scale coal mining of the Joban coalfield in Iwaki, Japan (Figure 1) has been excavated down to the depth of 1200m below sea level during the operational periods of the past 100 years since the early Meiji era (1870's). Although the mine itself has been abandoned in 1970's, the mine company still keeps the geological information of the mine, the groundwater quality and the flow rate records of the mine drift seepage points, and their chemical and volumetric change data accompanying the coalmine development activities. The evaluation and interpretation of these mine operational time records are quite important because the condition of the groundwater disturbance resulting from large-scale and long-term underground excavations can be obtained in the Tertiary sedimentary bedrock aquifer with relatively low permeability.

In this report, the elucidation of the actual condition of seawater intrusion in the coastal sedimentary bedrock aquifer is discussed based on the geology and the groundwater chemistry collected during the coalfield operational periods.

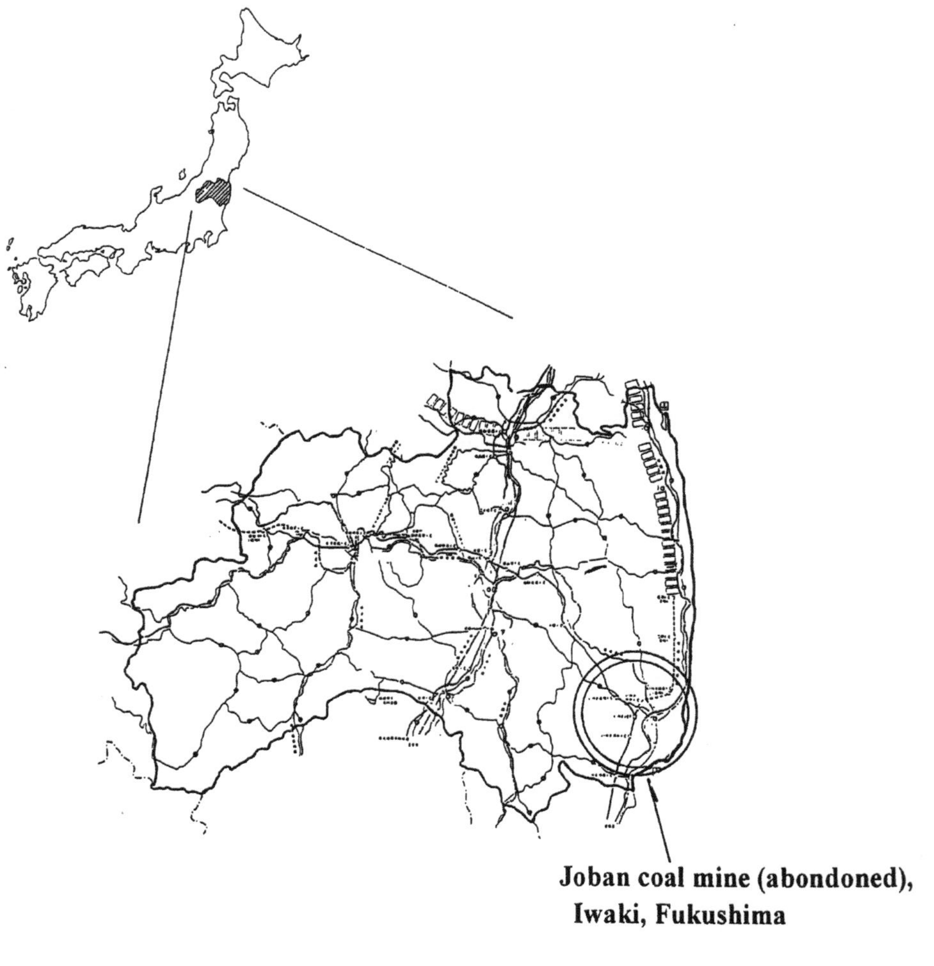

Figure 1. Location of the Joban coalfield.

2. THE GEOLOGY OF THE JOBAN COAL FIELD AREA AND THE EXCAVATION DEVELOPMENT OF MINE DRIFT SYSTEM

The Joban coalfield area is located at the Eastern foothills of Abukuma-mountain area in which there exists several fault systems with the direction of NW-SE. The area consists of a hill, lowlands, and a coastal plain. In this area, there are several geological reports on the sedimentary structure including the distribution of coal resources. The most recent

Table 1 Generalized geological column of the Joban coalfield area (Yanagisawa, *et al*, 1989).

Age		Group		Formation	Thickness (m)	Columnar section	Lithology
Quaternary				Alluvium			Gravel, sand and silt
Quaternary				Terrace deposits			Gravel, sand and silt
Pliocene	Late	Sendai G.	Upper	Tomioka F.	100		Sandy mudstone
Pliocene	Early	Sendai G.	Upper	Hirono F.	220		Sandy mudstone Coarse-grained sandstone
Miocene	Late	Sendai G.	Lower	Yotsukura F.	50		Sandy mudstone Conglomerate
Miocene	Late	Taga G.		Minamiisowaki F.	40		Sandy mudstone
Miocene	Middle	Taga G.					
Miocene	Early	Takaku G.		Shimotakaku F.	130		Mudstone
Miocene	Early	Takaku G.		Numanouchi F.	70		Fine-grained sandstone
Miocene	Early	Takaku G.		Kamitakaku F.	70		Coarse-grained sandstone
Miocene	Early	Shirado G.		Nakayama F.	170		Conglomerate, tuffaceous sandstone, mudstone and tuff
Miocene	Early	Yunagaya Group		Taira F.	300		Cross-bedded sandstone Mudstone Sandstone and tuff breccia
Miocene	Early	Yunagaya Group		Kamenoo F.	100		Thin-bedded mudstone
Miocene	Early	Yunagaya Group		Mizunoya F.	100		Sandstone and mudstone
Miocene	Early	Yunagaya Group		Goyasu F.	200		Sandstone
Miocene	Early	Yunagaya Group		Taki F.	150		Sandstone, mudstone lignite and tuff
Early Oligocene		Shiramizu G.		Shirasaka F.	150		Gray mudstone
Early Oligocene		Shiramizu G.		Asagai F.	100		Fine-grained sandstone
Early Oligocene		Shiramizu G.		Iwaki F.	250		Conglomerate, sandstone mudstone and coal
Late Cretaceous		Futaba G.		Tamayama F.	100		Coarse-grained sandstone
Late Cretaceous		Futaba G.		Kasamatsu F.	140		Sandstone and mudstone
Late Cretaceous		Futaba G.		Ashizawa F.	200		Fine-grained sandstone
							Amphibolite Sandstone limestone and slate / Granitic rocks

geological study based on the micro-fossils has been done by Yanagisawa (1989), and the resulted generalized geological column of the Joban area is shown in Table 1. According to these studies, many sedimentary rocks from the Late Cretaceous system, Early Oligocene series, Miocene to Late Pliocene series, and the Quaternary series are distributed over the Abukuma meta-sedimentary rock system and the Abukuma granitic pulton. The Iwaki formation, which constitutes the lower most part of the Tertiary system, consists of alternated strata of marine sandstone, gravel stone and several coal layers from the Palaeogene period. It outcrops at about 10km inland from the present coastline, and it was reported that the development of the Joban coalmine was started from this outcrop area. The underground drift excavation of the coal mine had been carried out along this Iwaki coal formation, with the general strike area toward the NW-SE and with a dip of 10 degrees to SE reaching a final depth of 1000 to 1200 m, ending about 4km inland from the present coastline (Figures 2 and 3).

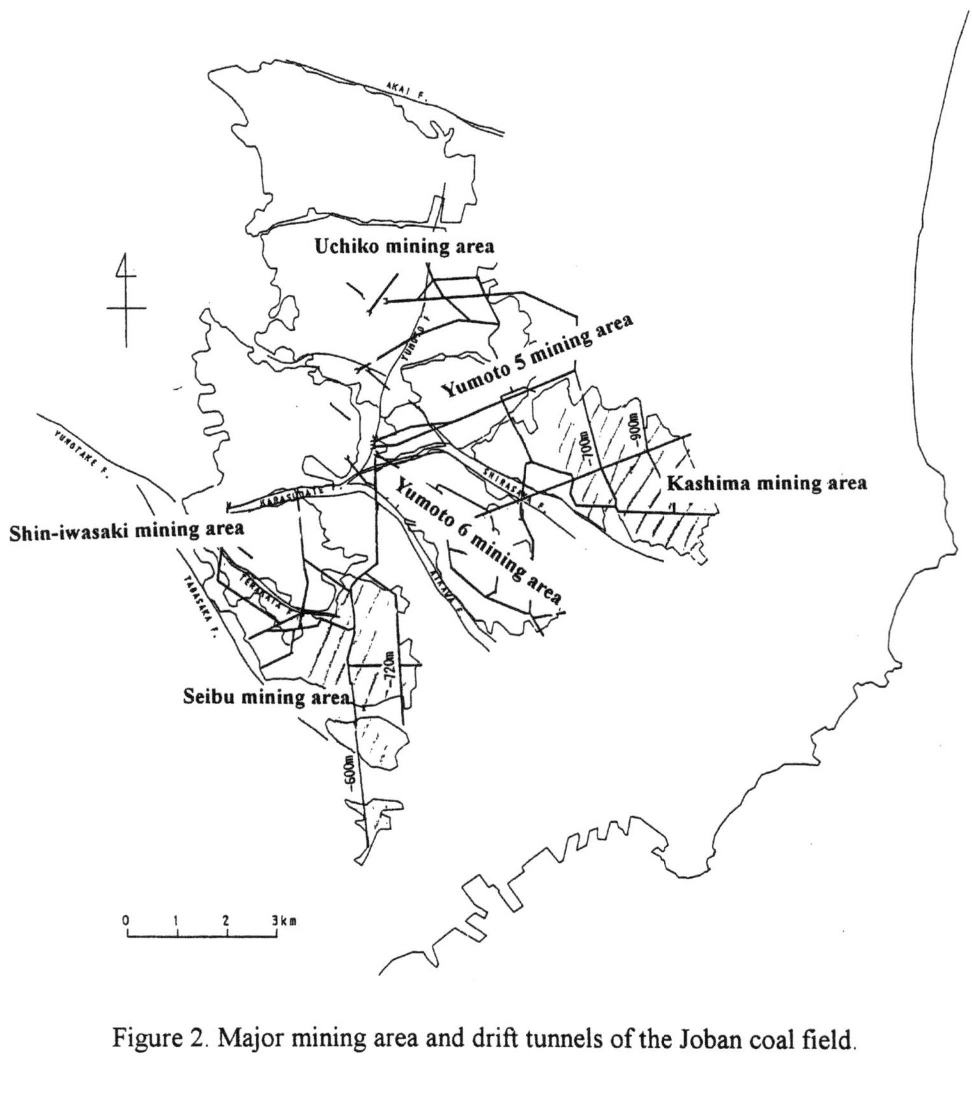

Figure 2. Major mining area and drift tunnels of the Joban coal field.

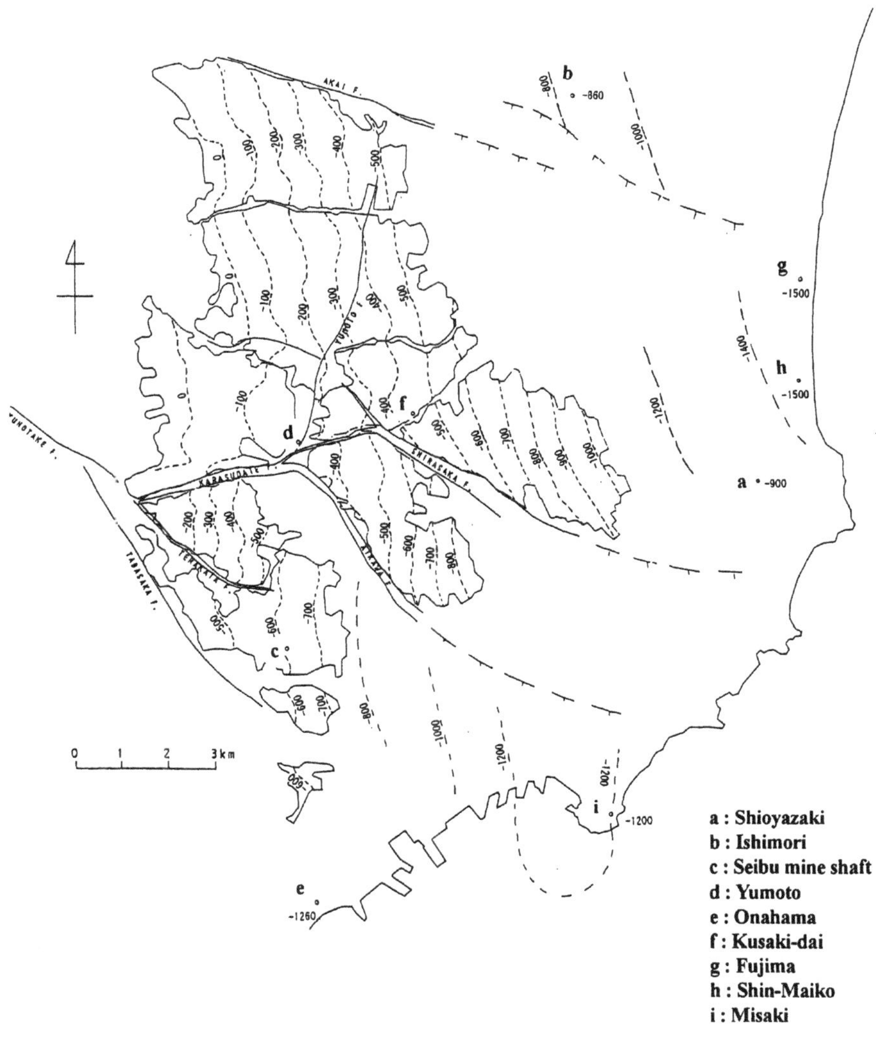

Figure 3. The structural contour map of Iwaki coal bed formation in the Joban coalfield area and the location of the hot spring around the mine. (Figure beside the hot spring point denotes the well depth)

It became difficult for the coalmine operation to excavate deeper than the depth of 1200m because of high CO gas content and the high ground temperature at the bottom drift of the final mining stage. The mine operation was shutdown in 1971 (Chishitukiso -kogyo, 1998).

Offshore drillings for the geological and geophysical surveys were conducted for natural gas and oil resources exploitation during the 1970's. According to these surveys, the Tertiary sedimentary formation also continues to the offshore Joban area. The long axis of the fold runs sub–parallel to that of the Honshu Island in an extensional direction, and its stratum thickness becomes increasingly thicker towards the central part of Honshu (Komatsu, 1979).

3. WATER QUALITY CHARACTERISTICS OF GROUNDWATER IN THE JOBAN COAL MINE

In the Joban coal mine drift, two types of groundwater seepage are observed by the mining period's experiences; (1) type-A seepage which flows out to the drift tunnel from the upper ceiling strata of the coal formation, and (2) type-B seepage which flows out into the drift tunnel through boreholes and bedrock cracks from the lower bedrock board beneath the coal formation (Figure 4). Type-A seepage has been relatively higher than type-B in relation to ion concentrations, but generally has decreased over a few months to a few years after drift excavation, while type-B seepage has flowed out continuously to the drift tunnel. The water quality of these two types of groundwater seepage differs remarkably; this is probably affected by the difference of the groundwater flow system around the mine drift. In the following section, the continuous flowing type-B groundwater seepage has been mainly discussed for their chemical characteristics.

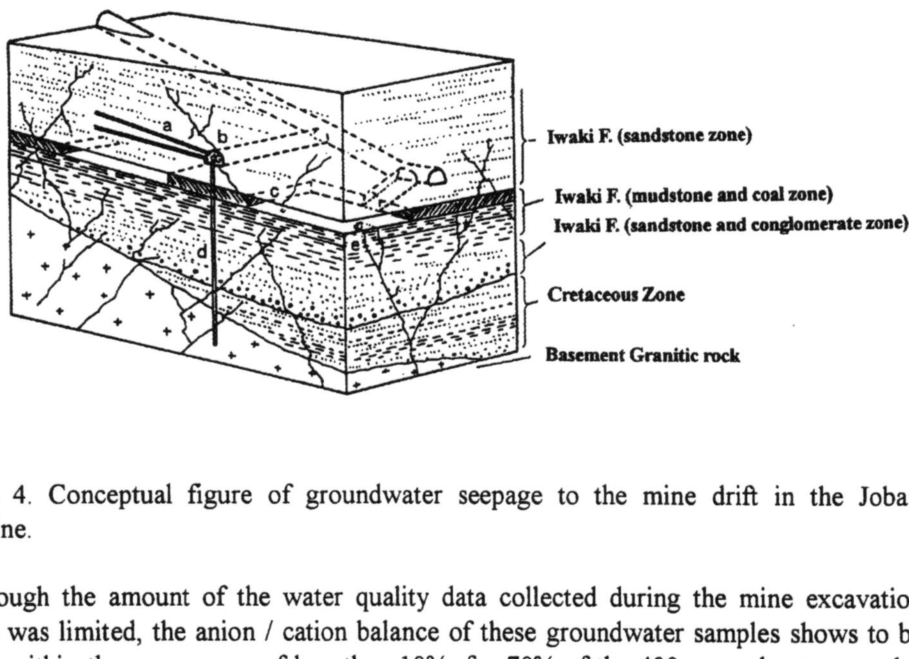

Figure 4. Conceptual figure of groundwater seepage to the mine drift in the Joban coalmine.

Though the amount of the water quality data collected during the mine excavation period was limited, the anion / cation balance of these groundwater samples shows to be settled within the error range of less than 10%, for 70% of the 400-groundwater samples analyzed for seven ion components. Based on above analytical results, the accuracy of the water quality data has enough reliability to compare with the present chemical data of surrounding hot springs. Based on this analytical accuracy, the water quality characteristic of the groundwater seepage in the drift was examined using about 1200 groundwater samples in which includes in the data of single or semi-single ion component analysis remaining until present. The result shows clear water quality differences among the main drift development areas.

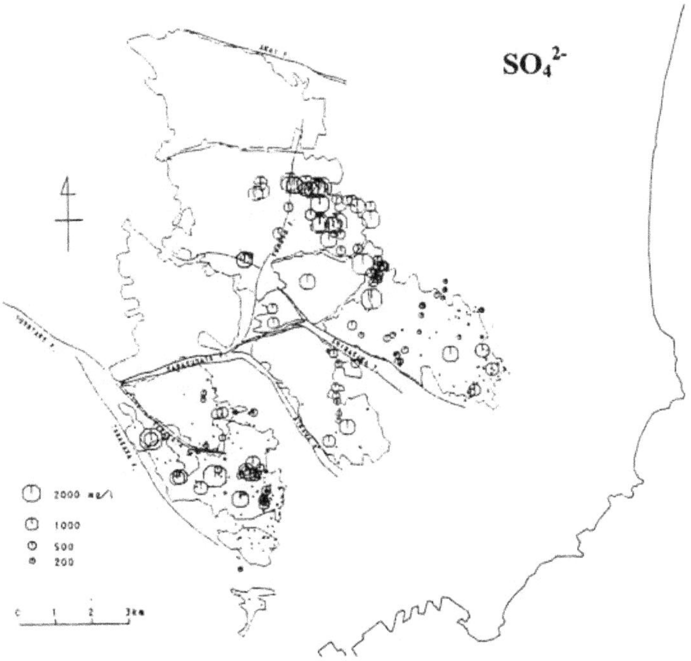

Figure 5. SO_4 ion concentration distribution map of groundwater seepage in the mine drift.

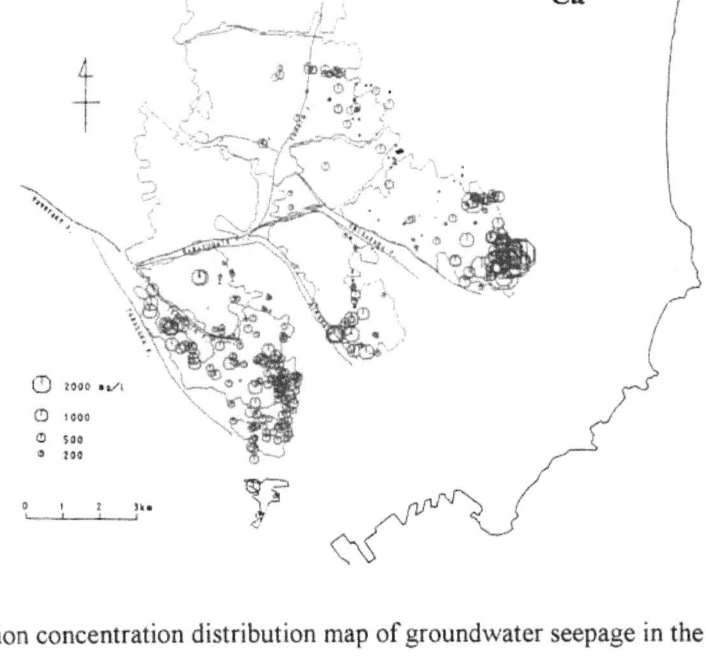

Figure 6. Ca ion concentration distribution map of groundwater seepage in the mine drift.

The groundwater quality in the mine, based on the distribution of Cl, SO_4, and Ca ion concentrations of the groundwater seepage (see Figures 10, 5, and 6) was classified into three types shown as follows.

(1) A part of the groundwater seepage of the Yumoto 5 mining area and the Uchiko mining area, where a SO_4 ion content of 2000 mg/l or higher but little or no Cl ion content. This relatively high SO_4 content in spite of a relatively shallow depth (200-600m) is different from that of the mining areas of Kashima, Yumoto 6, and Seibu that are deeper than 500 m and have relatively high Cl and SO_4 contents.
(2) Seepage from upper layer of the coal bed formation at the Kashima mining area, the Yumoto 6 mining area, and the Shin-Iwasaki mining area, shows Ca-Cl type groundwater with a high (5000-10000 mg/l) Cl ion content (these are not shown in Figures).
(3) Seepage from bottom layers of the coal bed formation at the Kashima mining area, the Seibu mining area and the Yumoto 6 mining area have a Cl concentration of about 1000-2000 mg/l.

As the type (1) groundwater can be characterized by less of the Cl ion component, the source of SO_4 ion is not the seawater or the fossil seawater. It is the contribution of the sulfur from the pyrite in the coal bed itself. On the other hand, a high Cl concentration with rich Ca components show the evidence of the fossil seawater, and thus the origin of the type (2) groundwater has a high possibility of the fossil seawater. As for the origin of type (3) groundwater, it is thought that the surface groundwater into the drift was mixed with the Cl ions of the original seawater because their Ca/Cl ratio and SO_4/Cl ratios are almost identical with that of the present seawater. In addition, the increase of Cl concentration with depth and the stable oxygen data for mine groundwater of this area (Takase, 1981) support that the origin of type (3) groundwater is seawater. This possible explanation will be discussed more in the next section.

The above groundwater types (1) and (2) are thought to reflect a relatively local-scale groundwater flow system, where the original formation groundwater has flowed into the drift by the effect of excavation. Whereas especially type (3) groundwater, it might be affected by the regional scale groundwater flow system including the location of the salt-fresh water interface zone in a coastal sedimentary bedrock aquifer.

4. ORIGIN OF Cl-IONS ESTIMATED FROM STABLE ISOTOPE RATIOS

According to the Cl ion, δD, $\delta^{18}O$, and ^{14}C content in the drift groundwater seepage at the Seibu mining area (Takase, 1981), it is interpreted as shallow groundwater with the surface origin and the seawater being mixed. In addition, these mixed groundwaters have had thousands of years to tens of thousands of years resulting in ^{14}C content in the water. Although the number of isotope data studied by Takase was limited, the environmental isotope information for the groundwater seepage in the mine drift has not reported other than in this study. As the coal mine has already abandoned at present, it is impossible to sample the water from the mine drift itself. It is also impossible to reconsider the water quality classification of the mines groundwater based on the newly sampled environmental isotope information.

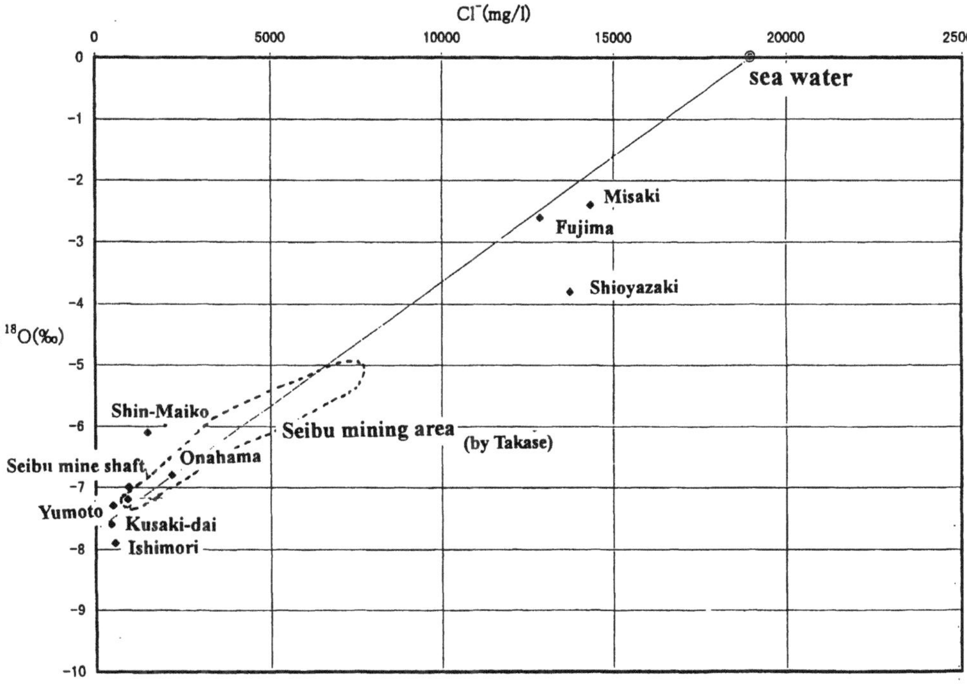

Figure 7. The relation between Cl ion content and stable oxygen isotope ratio of the groundwater seepage in the mine drift and hot spring waters around the mine.

To compensate this, the subsequent nine samples of the hot spring water around the abandoned coal field (Figure 3) were taken in February, 1996 and January, 1997 to examine the relationship between mine groundwater and the hot spring water especially for the origin of Cl ion. During this sampling, the groundwater from different depths of the abandoned mine shaft (the Seibu mining shaft that has already filled by the recovered groundwater) was also sampled as a representative groundwater sample of the mine. The inorganic water chemistry and the environmental tritium and stable isotopes were analyzed for the sampled waters. Figure 7 shows the relationship between the Cl ion content and the stable oxygen isotope ratio for the groundwater seepage in the mine drift by Takase (1981) and the nine hot spring water samples including the shaft groundwater by this study. According to this figure, they are distributed along the line which connects the average local shallow groundwater (having around – 7.5 ‰ of $\delta^{18}O$ and 10 to 20 mg/l of Cl) and the seawater. This distribution suggests that the origin of Cl ion is the seawater. However, there are two kinds of possible seawater; the present seawater and the fossil seawater depending on the time when the seawater taken into the sedimentary stratum.

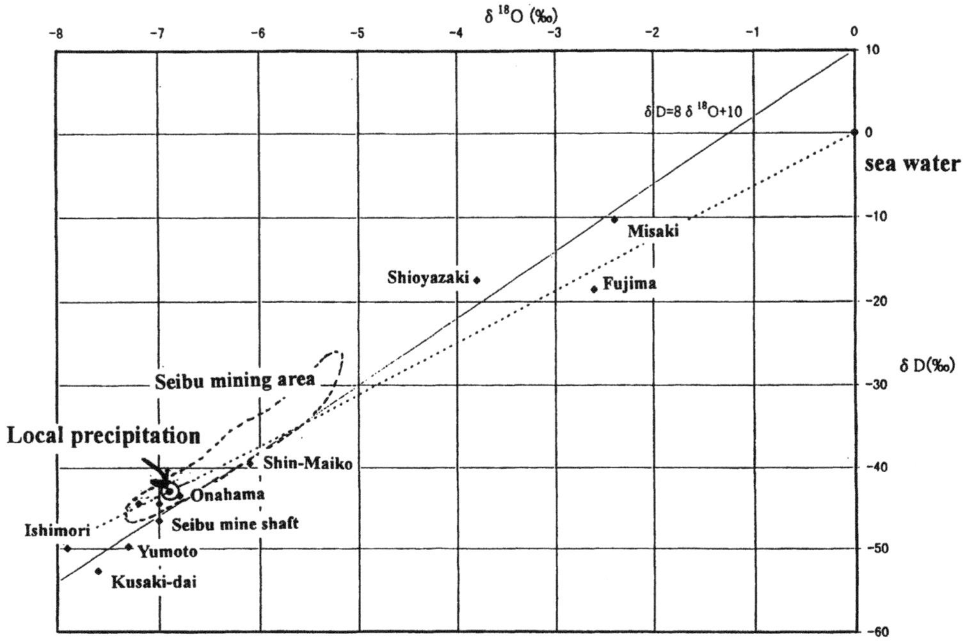

Figure 8. Delta diagram of groundwater seepage in the mine drift and the hot spring waters around the mine.

The previous study of the mine groundwater by Takase (1981) did not mention the difference between these two kinds of seawater. Figure 8 shows the delta diagram of the hot spring water in the study area, a seepage groundwater in drift, and local precipitation (annual isotopic mean of the precipitation at Mito (locating about 80 km SSW of the study area) being δD: - 43.0 ‰, ^{18}O: - 6.9 ‰). According to Figure 8, two typical groups can be classified; (1) the group very close to the local precipitation and (2) the group plotted along the mixing line connecting the local precipitation and the seawater (δD: 0 ‰, δ^{18}O: 0 ‰).

Generally, the isotopic oxygen content of the fossil seawater is known to be heavier than that of present seawater (ex. stable oxygen ratio with +2 to +20 ‰). This is because of the isotopic exchange within the stratum during a long period of being trapped in the formation. Although the number of the isotopic samples for the mines groundwater in the Joban coalmine is limited, the above-mentioned isotopic characteristics with the isotopic oxygen shift has not yet confirmed the seawater plot in the delta diagram of Figure 8. By this evidence, it is reasonable to consider that the recent seawater could be the origin of Cl ion in the groundwater of the study area, which is characterized by the relatively higher Cl content in the Seibu mining area, in the coalfield and the hot spring water around the mine.

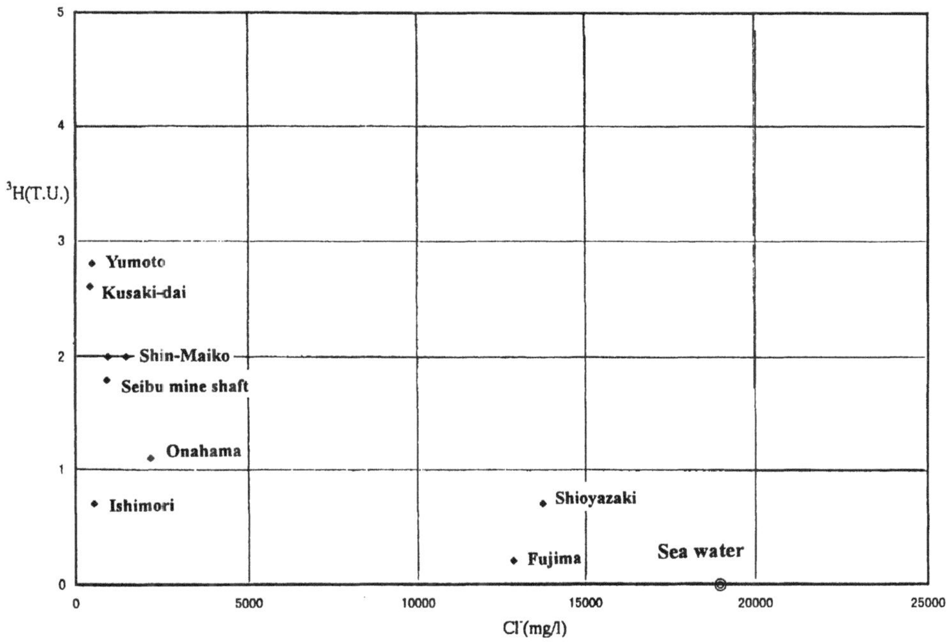

Figure 9. The relation between Cl ion and tritium concentration of the groundwater seepage in the mine drift and the hot spring waters around the mine.

5. THE ACTUAL CONDITION AND A SPATIAL DISTRIBUTION OF SALT-FRESHWATER MIXTURE

In the previous section, it has confirmed that the chemical and isotopic components of the groundwater seepage in the mine drift and the hot spring water around the Joban coalfield show the mixture components of the groundwater. The shallow groundwater recharged by the modern precipitation has been mixed with the salt water having a composition close to the present seawater. In this section, the actual condition of these mixing processes will be considered.

If the stable isotope ratio of the shallow groundwater in the study area is – 7.5 ‰ ($\delta^{18}O$) and the Cl ion content is nearly zero, it is possible to confirm the mixing ratio of the seawater (having 0 ‰ for $\delta^{18}O$, 18980 mg/l for Cl content) by using a two component mixture model for the hot spring water and the groundwater seepage in the mine drift.

As shown in Figure 9, the tritium concentration in both the hot spring water as well as in the mineshaft groundwater has a remarkable decreasing tendency when Cl levels increase. It can be derived from Figure 9 that the tritium concentration in the seawater extrapolated to be 0 T.U.

It is possible to estimate the tritium concentration in the surface shallow groundwater by using the measured tritium concentration in the hot spring water around the mine and the mixing ratio of the seawater obtained from the above mentioned two componet mixture model of Cl and stable isotope ratio. The apparent age of a mixed shallow groundwater can be evaluated as about 10 - 40 years by the simple radioactive decay model. While, the apparent age of the seawater should be more than 50 years because their estimated tritium concentration is lower than 0 T.U.

The Cl contents distribution of the hot spring water around the Joban coalmine obtained by this study and that of the type (3) groundwater seepage are shown in the composite figure as Figure 10. In this figure, the Cl concentration of about 1000-3000 mg/l at the Seibu mining area and at the Yumoto 6 mining area was considered the result of the mixture of the seawater and the shallow groundwater.

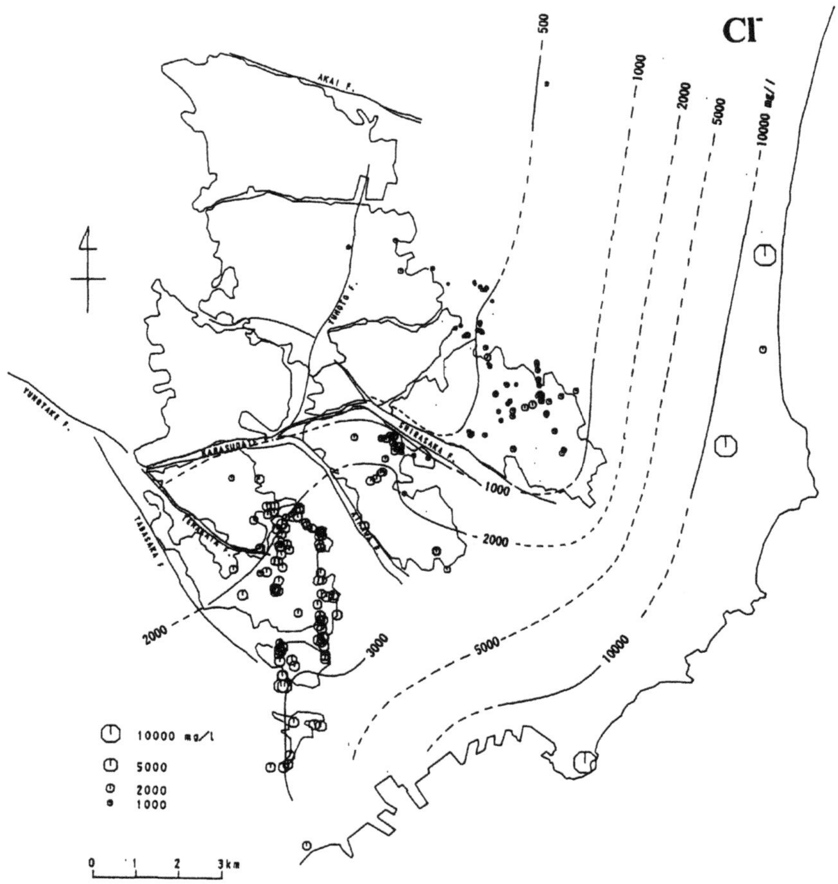

Figure 10. Cl ion content distribution map of groundwater seepage in the mine drift and the hot spring waters around the mine.

The Cl concentration in the hot spring water, such as the Taiyo-no-sato hot spring, the Yumoto hot spring, the Kusaki-dai hot spring, and the groundwater in the Seibu mineshaft (locations of these hot springs are shown in Figure 3), shows very close Cl contents with that of the groundwater seepage in the coalmine drift. This means that both the groundwater seepage in the coal mine drift and the hot spring water have a similar mixing process to the present seawater and the shallow groundwater.

Figure 10 shows that the Cl ion content clearly increases toward the coast and it can be explained by this Cl ion content distribution that the salt-fresh-water interface zone exists in the inland part of the Tertiary sedimentary bedrock. It is possible to consider that this is the result of the Joban coalmine excavation down to the depth of 1000m or more during the past 100 years. In the case that the Cl ion content in the seawater is 19000 mg/l, a contour line of a Cl ion content with 10000 mg/l can be considered as the 50 % seawater mixing zone. Moreover, if we consider the stable dip of 10 degree for the coal bed and their upper and lower formations, the seawater side of the salt-fresh-water interface zone with 19000 mg/l Cl content, will exist near the depth of about 1100-1800 m below surface. This is confirmed by the Cl content and depth relations along the coal bed formations, where the horizontal permeability of these sedimentary bedrock formations is almost uniform with a stable dip angle.

However, it is necessary to examine the more exact permeability distribution that reflects the actual geological distribution for understanding the precise groundwater flow system around the salt-fresh-water interface zone.

6. CONCLUSION

Generally, the interface of the fresh groundwater and seawater is pushed out toward the seaside according to the relatively higher groundwater potential distribution of the inland fresh groundwater behind the coastal area (Figure 11; Seki *et al.*, 1980, Minakami, 1996, etc.).

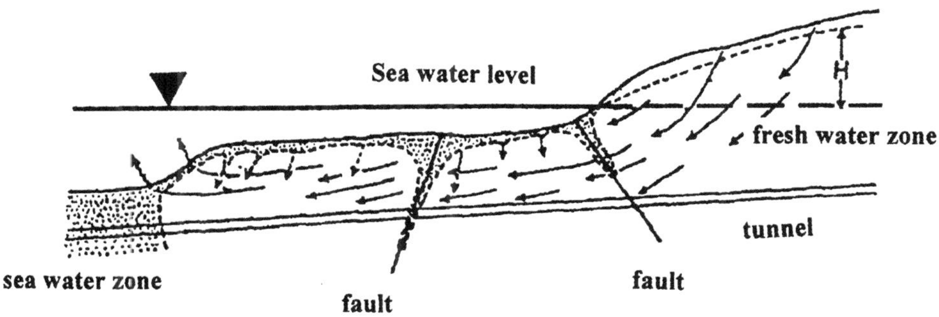

Figure 11. Schematic figure of the groundwater flow along Seikan undersea tunnel (Seki, Y., *et.al.*, 1980).

When we consider the stage of marine transgression and regression accompanying the expansion and reduction of the continental glacier development during the Quaternary period, the present is in a relative transgression stage. This started about 6000 years ago. If the salt-fresh-water interface formed at the previous marine regression stage (approximately 20,000 years ago), and remains until present, it should exist beyond the present coastline to the offshore side. This could not explain the situation where the salt-fresh-water interface exists farther inland than the present coastline, which has confirmed by the present study.

The permeability of the Iwaki coal bed formation, which is an alternated stratum of the sandstone and mudstone of a Tertiary sedimentary formation layer, is about an average of 10^{-5} cm/sec. In spite of these low permeability characteristics of the sedimentary bedrock, this study revealed that a large-scale groundwater disturbance has created a salt-fresh-water interface zone toward the inland. It is reasonable to consider that the cause of this driving force was the operation of the Joban coalfield mine excavated down to a depth of 1200m during the past 100 years.

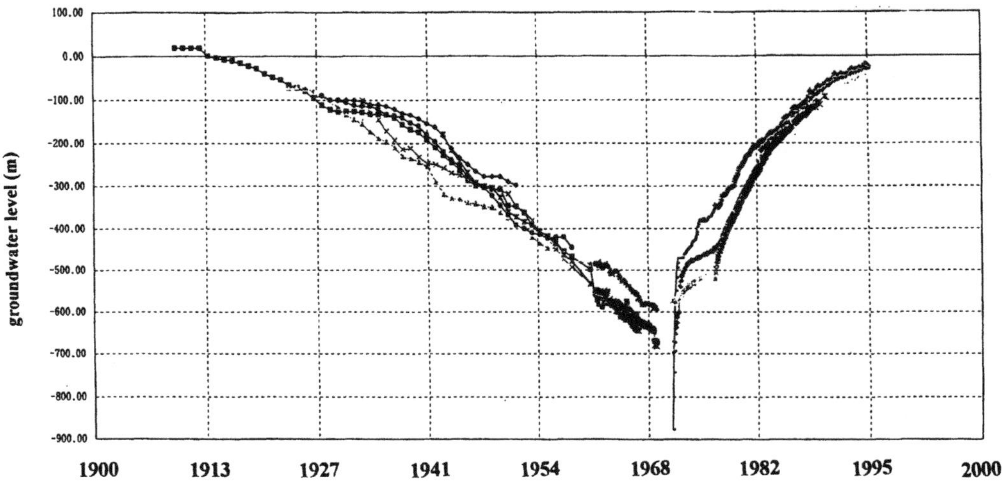

Figure 12. The groundwater drawdown and its recovery caused by the Joban coalfield mining excavation and abandonment.

Figure 12 shows the existence of a large-scale groundwater drawdown and its recovery caused by the mine operation and shutdown. This kind of long-term and large-scale drawdown of the groundwater level must influence the groundwater flow system even in the low permeability sedimentary bedrock, where the groundwater flow system should be very stagnant. As a result, this drawdown creates the new artificial groundwater flow toward the deep mine excavated area causing the salt-fresh-water interface to move inland.

The salt-fresh-water interface of the low permeability sedimentary bedrock has not been precisely studied by this report, because this study was mainly based on the limited data during mine operational time. In addition, the exact age of the mixed seawater is unknown. The water chemistry of the coastal sedimentary bedrock groundwater may be affected by the water in the pore occupied by the original seawater included during the deposition of the undersea formation. We would like to leave it to the future environmental isotope studies using much older groundwater-age tracers to distinguish the difference of these pore-waters in the sedimentary bedrock.

References

Chishitukiso-kogyo, 1998. Heat condition in Joban coals mine (*in Japanese*). Taira electronic printing office .366p.

Hayashi, T., *et. al.*, 1999. Characteristics of water chemistry of inland water and submarine discharged groundwater in Rishiri island, Northern Japan (*in Japanese*). *Journal of Japanese Association of Hydrological Sciences,* **29**, 123-138.

Komatsu N., 1979. The deposition basin in Joban and the northing offing (*in Japanese*). *Japanese Association for Petroleum Technology*, **44**, 36-39.

Marui, A. and Yasuhara,M.,1999. Studies on groundwater flow systems with a saltwater-freshwater interface (*in Japanese*). *Journal of Japanese Association of Hydrological Sciences*, **29**, 1-12.

Minakami, M.,1996. Geochemistry of the Seikan Tunnel groundwater seepage in drift (*in Japanese*). *Chikyu monthly Special issue* No **16**.159-164.

Seki, Y., *et al.*, 1980. The interaction between Miocene volcanogenic rocks and the sea water-meteoric water mixtures in near coast undersea part of the Seikan Tunnel, Japan (*in Japanese*). *Saitama University groundwater hydraulics research center report* No.**1,** 1-123.

Takase, K.,1981. The origin of the hot spring and their water chemistry –example of the Joban coal mine seepage hot spring groundwater in the drift (*in Japanese*). *Applied geology*, 422-428.

Yanagisawa Y., *et al.*,1989. Tertiary bio-stratigraphy and subsurface geology of the Futaba district, Joban Coalfield, northeast Japan (*in Japanese*). *Bulletin of the Geological Survey of Japan.* **40**, 405-467.

Land and Marine Hydrogeology
Edited by M. Taniguchi, K. Wang and T. Gamo

Natural tracing in karst aquifers

Michel Monnin and Michel Bakalowicz
Centre National de la Recherche Scientifique. Hydrosciences, UMR5569
Place E. Bataillon, 34095-MONTPELLIER, France

Porous and fissure aquifers show statistical homogeneity of their physical and hydraulic characteristics on a scale ranging from tens to several hundreds of meters. Their groundwater resources can actually be explored by a simple approach, i.e. defining the aquifer geometry from geological data, and determining local hydraulic parameters from pumping tests; finally, local data are generalized to the whole aquifer through regionalizing techniques.

On the contrary karst generating processes create a peculiar void heterogeneity: voids may reach several meters in diameter and several kilometers in length. These voids are organized in a hierarchized network from the surface to an often unique spring: this is the conduit or drainage network. In this context, karst hydrogeologists must adopt a different approach to characterize a karst aquifer, its resources and storage, i.e. via an accurate description of the void organization and an analysis of its overall behavior (or functioning) and that of its different parts or "organs". By such an approach, a karst aquifer is considered as a living organism composed of different types of organs interlinked by functional relationships.

Therefore techniques are used for describing the aquifer in a bi- or tri-dimensional space (geology, geophysics) and for characterizing its functioning (hydrodynamics, natural tracing, hydrological balance). In addition, data from these techniques are interpreted so as to be able to propose a diagnosis, i.e. for building a conceptual model of the studied aquifer. At last, as in medicine, the conceptual model can be assessed from localized tests, such as artificial tracing or diver's exploration for borehole positioning and pumping tests. Methods for interpreting tracing and pumping tests must obviously be adapted to karst specificity, since they cannot be based on classical models whose basic assumptions are never verified in karstic medium.

1. HISTORICAL AND GEOGRAPHICAL BACKGROUND

Originally, "Karst" is the German name of the limestone plateaus situated in the north-western part of the Balkan peninsula, between Carniola and Istria. This region constitutes the hinterland of Trieste, in the northern Italy. Karst derives from the slav word, Krs or Kras, (in Italian Carso), which comes directly from the indo-european root kr (or ker) whose meaning is "mountain".

However, interest in karst phenomena can be traced back to ancient times and though investigation on karst were mentionned in Chinese documents (Tsin period, 316 - 265 B.C.), the Mediterranean basin is the craddle of karstic studies. "Ancient Greek and Roman philosophers made the first known contributions to our scientific ideas on karst, as well as contributing to a mythology as well, like the River Styx, lives on in the place names given by cavers and others. Pfeiffer [1963] identified five epochs in the development of ideas about karst grounwaters, in the interval 600 - 400 BC until the early 20th century. Thales (624 - 548? BC), Aristotles (385 - 322 BC) and Lucretius (96 - 45 BC) formulated concepts in the nature of water circulation. Flavius (1st century AD) described the first known attemps at karst water tracing in the River Jordan basin [Milanovic, 1981]. Pausanias (2nd century AD) also reported experiments that were interpreted as proving the connection between a stream-sink beside Lake Stymphalia and Erasinos spring [Burdon and Papakis, 1963]. The conceptual understanding of hydrology established by Greek and Roman scholars remained the basis of the subject until the 17th century, when Perrault (1608-70), Mariotte (1620-84) and Halley (1656 - 1742) commenced its transformation into a quantitative science, showing the relationship between evaporation, infiltration and streamflow [Biswas, 1970]. By the end of the 18th century, the role of carbonic acid in the dissolution of limestone was understood [Hutton, 1795]. Experimental work on carbonate solution in water followed a few decades later." [Ford and Williams, 1989]

More recently, the Karst, which was part of the XIXth century Austro-Hungarian empire, has been systematically put under study since 1860 by Austrian and Slav academics and engineers, in order to better exploit it. The economy of the area was mainly based upon agriculture and, hence, water was a major concern, either because of the frequent disappearing of surface water into swallow-holes, depriving communities of their water supplies, or because of the seasonals floods inundating the closed karstic plains (dolines and poljés), gravely perturbating agricultural processes. Accordingly, the first works were devoted to the study of the circulation of underground and surface waters. Engineers played then an essential role in the analysis of the phenomena in the view of remediating the encountered problems. Von Mojsisovics [1880] was probably the first investigator who considered karst as an individual and peculiar physiographic whole. Soon after and following him, German authors have extended the meaning of karst to all the calcareous massifs of Croatia and Herzegovina [e.g. Cvijic, 1893; Penck, 1900, 1903; Grund, 1903] and of Russia [Kruber, 1900]. The German terminology is related to landscapes ("Karst-Erscheinungen" or "Karstphänomen") as well as to water flowing ("Karsthydrographie"). Italians did so as well [e.g. Del Zanna, 1899]. In France, de Lapparent [1883] suggested to generalize the meaning of the term.

Nowadays, karst is defined as an unity of surface and subterranean landscape resulting from the dissolving of limestone of dolomies by groundwater rendered acid by carbon dioxide. In a wider sense, the word karst refers also to any comparable structure that develops onto saline rocks (haline, gypse, anhydrite), also called "pseudo-karst".

Several important concept are associated with karst. Among them, the "karst system" concept plays a prominent role in hydrogeology, for a specific methodology has derived from it, which is closer to that of surface hydrology than that of classical hydrogeology. This expression was introduced for the first time by Mangin [1974] who defines the "karstic system" as "the entirity of underground flows of the karstic type which organize themselves to constitute a draining unit". Accordingly, the karstic system corresponds to the feeding hydrogeological basin of a karstic spring or of a group of kartic springs connected to one another. The karstic system can be constituted only by karstified carboneous formations; it is then a "authigenic karstic system" [Jakucs, 1977] according to the anglo-saxon nomenclature, or an "unarian system" [Mangin, 1978] according to the terminology of systemic analysis. When the karstic system also includes non-karstic formations, submitted to surface flows drained by swallow-holes, it is said "allogenic" or "binary".

Limestones and dolomites are widely spread over the World. Their outcrops cover about 12% of the emerged and ice-free land, but develop also below non karstic terrains. About 25 % of the world population is supplied with groundwater from karst aquifers. In Europe, about 35% of the land are karstic and 35 to 40% of domestic water supplies are of karstic origin. In the former Soviet Union (where up to 40 percent of the surface consists of carbonate and other soluble rocks) vast karst areas exist, as well as in the United States where these rocks comprise about 15 percent of the territory. Still in America, the Circum-Gulf Karst Belt encompasses Jamaica, Cuba, Puerto Rico, the Bahamas, the Yucatan Peninsula and part of the Sierra Madre in Mexico, and of Guatemala. In Africa the most abundant karstic areas are Israel, Tunisia and Lybia in the North, and Transval in the South. In Asia, karst can be found from the East, Turkey and Saudi Arabia, to the West, Vietnam (Shon La and Biu Chi Tien regions) and China (regions of Guangxi, Chuang, Guangdong, Guizhou, and Yunnan). The Nullarbor platform in Southern Australia is one of the largest karst region in the world. In short, karst aquifers are widely distributed over the world.

2. THE EXPERIMENTAL APPROACH OF AQUIFERS

Groundwater has to be explored for assessing its resources for a sustainable development either in order to exploit them or in order to protect them from man polluting activities. Their exploration may be led at two different scales: at a regional scale, for assessing the effects of local withdrawals or of polluting activities on the whole aquifer, and at a local scale for the designing and the implementation of water-catchment works.

Traditionally, hydrogeologists use an efficient and well known approach developed for the exploration of porous aquifers. As a first step, the *regional approach* consists in a geological description of the aquifer and of its limits and boundary conditions, and in an assessment of its regional hydraulic characteristics from its physical characteristics (porosity, permeability). The geological description, consisting in direct observations (field works, boreholes) and indirect observations (geophysics), attempts to propose a geological conceptual three-dimension model. From this description, the best sites for drilling wells, or for implanting potentially polluting activities are chosen in order to get the best benefit and to manage the resource in a sustainable way.

Afterwards, the *local approach* is worked up in order to assess the local effects liable to affect the aquifer (exploitation pumping rate, recharge rate, flow of pollutant, etc.). It consists in localized tests : pumping or well tests, infiltration and tracing tests.

Field studies have shown early that porous aquifers offer a statistical homogeneity of their physical and hydraulic characteristics on a scale ranging from tens to few hundreds of metres. Such an homogeneity is linked with the relatively narrow space variability in these characteristics, and corresponds to a hydraulic continuity in the whole saturated zone. Therefore, local information may be generalized to the whole aquifer through regionalizing techniques.

Modeling is then used at the above mentioned two space levels, in different ways. At the local level, models allow the calculation of flow and transport parameters (i.e. hydraulic conductivity, storage coefficient, dispersivity), from field tests, according to basic assumptions. Afterwards the computed values may be used for instance to calculate the exploitation flow rate of a well, or the local recharge. But these results are also common inputs in regional models to assess the long term effects of man activities or of climate changes, which cannot be experimented directly on the field.

This approach, convenient and efficient for exploring porous media, has been extended to fissured and fractured aquifers, with the assumptions that the hydraulic continuity is likely to exist in their saturated zone, as well as the statistical homogeneity, but at a larger scale level (several hundreds of metres) due to a wider variability of void dimensions. The fact that the voids may not be randomly distributed, but rather organized in fractures, is taken into account by introducing permeability in the form of a tensor.

Therefore, the exploration techniques should be perfected, in order to take into account the higher complexity of fracture medium. In particular, the fracture and fault distribution has to be minutely analyzed from field and from aerial photographs in order to assess the location of the most important groundwater flow paths. Geophysics may also contribute to this understanding. But, on the whole, the approaches of porous and of fracture media are not basically different and resort to the same exploration techniques.

Karst aquifers were and are still occasionally considered in the same way as fracture aquifers, assuming that karst features are simply enlargements of initial fractures. But, the classical approach commonly fails, particularly when karst features are abundant. In reality, within infiltration and saturated zones of carbonate aquifers, karst processes create a peculiar void heterogeneity. Voids may reach several metres in diameter and several kilometres in length and are organized in a hierarchized network from the surface to an often unique spring: the conduit or drainage network.

It can be pointed out that none of the above approaches are of stochastic essence. As a matter of fact the stochastic approach; which is occasionally used in surface hydrology studies, does not apply to the karst aquifers.

3. THE EXPERIMENTAL APPROACH OF KARST AQUIFERS

The existence of conduits and of large voids makes the karst medium very peculiar : it is the only aquifer Man may enter by means of natural ways and go through. Therefore, observations and measurements can be realized directly inside the aquifer. Since more than a century, speleological exploration is considered as a very efficient way to characterize flow conditions in karst. In few cases, the conduit network has fully developed and it functions like a sewage network draining a surface flow to a spring, through swallow holes. The direct exploration allows then to describe in a proper way the underground flow conditions. However most of the time, the conduit network is not explorable at all, or the groundwater flow observed in it is a small part only of that outflowing at the main spring.

As a matter of fact, karst aquifers must be distinguished once for all from porous and fracture aquifers and from pipe networks. The voids making karst an aquifer are not predetermined, but created by the groundwater flow itself, and develop in a relatively short time (several thousands to ten thousands years) from an initial distribution of fractures (the fracture pole) to a final organization of conduits (the karst conduit pole).

The final result, the karst structure (Fig. 1) , is the organization of voids in two

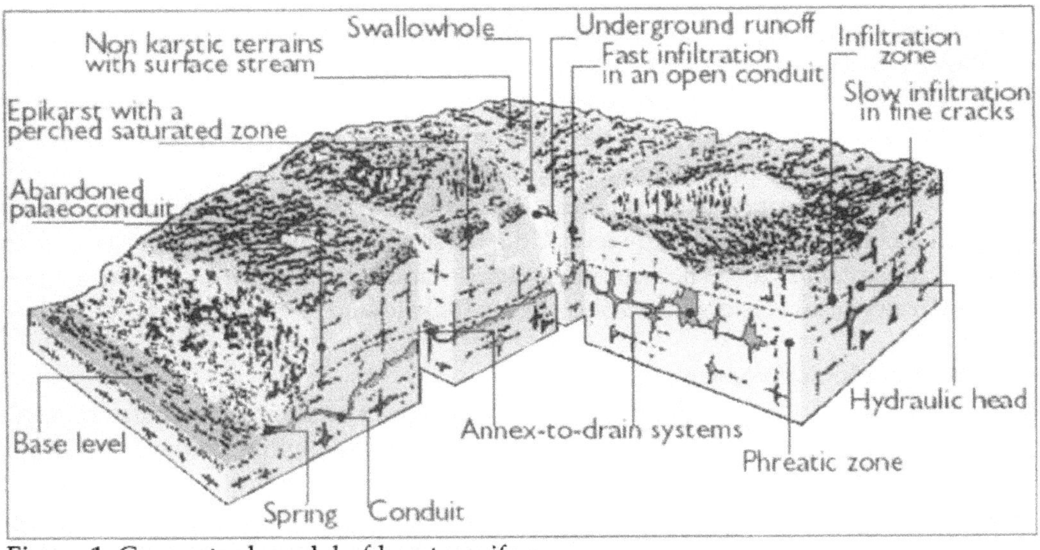

Figure 1. Conceptual model of karst aquifer

different functions : the network's transmissive or drainage function on the one hand and its links with storage parts, the storage function on the other hand. The drainage function is realized by the conduits ; groundwater storage occurs either in large karstic voids connected to conduits with high water head losses, or in blocks in which fine cracks and joints create a high porosity. Development of the karst structure within a carbonate rock formation depends on two independent sets of conditions :

- the *geological framework*, in which karst develops, characterized by the geometry of the carbonate reservoir and the relationships with non carbonate rocks, by the distribution of fractures, and the lithology (solubility and impurities of carbonate rocks);
- the *karst potential* [Bakalowicz, 1996], which determines the potential flux of dissolved carbonate flowing out of the aquifer and responsible for the intensity of karst processes.

Karst potential is composed of flow and solution conditions :

- underground flow conditions depend on rain and on non-karstic surfaces possibly drained by the karst aquifer. They are ruled by the regional water head, which is the difference in altitude between the recharge surface and the spring.
- solution conditions depend on CO_2 production, i.e. on climate and plant cover.

Geological or climatic events may change the functioning of the system, particularly by changing the position of its base level, on which lies its outlet. When the base level goes down (uplift, sea level lowering), the karst structure is abandoned and partly or totally included in the infiltration zone of the new system, which then belongs to the *fracture aquifer pole*. When the base level rises up (subsidence, sea level rise), the outlet is totally or partly blocked up and the karst structure is flooded and no longer functioning : it becomes a *non functioning karst system*.

Therefore, the characteristics of a karst aquifer are determined by :

- the initial geological framework,
- the state of development of the karst structure,
- the behaviour or functioning of the karst system.

However both approaches, classical and speleological, carried out alone usually give an incorrect or incomplete view of the karst aquifer and of its behavior and are finally unsuitable for characterizing karst aquifers and for assessing their groundwater resources and storage capacity, [Bakalowicz, 1997].

3.1. Methodological consequences

In this context, the karst hydrogeologist must adopt an approach similar to that of the physician or to that of the biologist examining living bodies : he has to characterize a karst aquifer, its resources and storage via an accurate description of the void organization and an analysis of its overall behavior and that of its different parts or organs. By such an approach, a karst aquifer is considered as a living body composed of different types of organs interlinked by functional relationships.

Nevertheless, contrary to the biologist, the karst hydrogeologist does not generally know a priori the "karst body" he has to study, because its hydraulic limits are not obvious.

As a consequence, prior to assessing groundwater resources of a karst aquifer and its possible storage capacity, the karst hydrogeologist generally has 1) to define the extension of the karst system, its limits and boundary conditions, and 2) to characterize the existence, state of development and functioning of the conduit network. Therefore, the approach of karst aquifers should develop [Crochet and Marsaud, 1997; Bakalowicz, 1999] following three steps :

1- *identification step* : rough delineation of the karst system and approximate definition of its limits and boundary conditions,
2- *demonstration step* : characterization of the existence, state of development and functioning of its karst structure, at the system level,
3- *assessment step* : design of a conceptual model of the system, assessment of the aquifer resources and storage capacity.

The demonstration step is essential, because its results will condition the methodology which will be put in operation in the third step. If the considered karst system does not show a typical karstic behaviour, either because it is close to the fracture aquifer pole, or a non functioning karst aquifer, it should be studied by means of classical methods (analysis of fracturing followed by the implementation of boreholes and piezometers, pumping and tracing tests) allow the exploration assessment of hydraulic characteristics from the knowledge of fracturing distribution and characteristics.

If the system shows a typical karst functioning, the structure approach is inoperative, because it is most of time unable to show the existence and the part played by the karst structure. Therefore the karst system must be studied by means of the functioning approach, which analyses the behavior of the system in order to determine the type and state of development of the karst structure and its hydraulic characteristics.

3.2. Exploration techniques

Therefore, as in the field of medicine, different complementary techniques are used to describe the aquifer in a bi- or tri-dimensional space (geology, geophysics) and its karstic development (geomorphology, speleology) and to characterize its functioning (hydrodynamics, natural tracing, hydrological balance). Three groups of techniques are then considered in the exploration of karst systems :

1) those describing the framework in which karst develops,
2) those describing the karst structure,
3) those analyzing the functioning of the system.

Data and information provided by all these techniques are eventually interpreted so as to be able to propose a diagnosis, i.e. for building a conceptual model of the system. At last, still as in medicine, the conceptual model can be assessed from localized tests, such as artificial tracing or diver's exploration.

As already said, the exploration of the karst structure is the domain of both speleologists and geomorphologists. Speleologists explore, describe and map caves. Geomorphologists describe, map and analyse all the karst landforms both at the surface and underground. Geomorphic analysis of karst landscapes are generally oriented in order to point out the genesis of the forms and the history of karst phases [Nicod, 1972 ; Sweeting, 1973 ; Ford and Williams, 1989]. The hydraulic functioning is considered, when distinguishing the landforms as "input forms" (swallow holes, dolines), "output forms" (springs, overflow conduits), "underground rivers" (conduits), or when considering [Renault, 1967] the shape of the local form, as "phreatic" (assumed totally filled with water during its development, not necessarily formed in the phreatic zone) or "vadose" (assumed partly filled with water.

However, caves and surface karst landforms alone do not represent the complete karst structure of a karst system. Input forms and conduits characterize only forms generally related to the transmissive function of the aquifer. The storage parts of the aquifer are known in exceptionnal cases only, when they correspond to wide open voids [Mangin, 1975]. Moreover, explored conduits are often not active : they are remains of ancient karst structure abandoned when the base level went down. Input forms may also be inherited from past karst phases and may not function as input forms in the present day structure. At last, many karst aquifers do not present any karst landform, either at surface level, when it is coated by thick sediments, or underground, when karst conduits are not enough enlarged or in the absence of caving exploration.

Geophysical techniques are commonly considered as very efficient to locate caves or conduits. Literature provides always the same few examples. But actual field works demonstrate the great difficulties one encounters for making the natural voids to come into sight. Benderitter, in a recent literature analysis [1997] showed that electrical and electromagnetic methods are not very efficient, but that microseismics and, overall, microgravimetry may be very efficient if the voids are large enough. However their localization by these means is not precise.

The electromagnetic positioning method allows a very accurate location of the cavities when an electromagnetic source (such as a powered solenoid) can be put inside; for example by a caver or a diver. The method, known since the 1970's, is presently commonly used thanks to electronics and computers. In France, Bardot and Co developed a very efficient electronic system which is successfully used for positioning boreholes. The precision, some centimeters down to 300 m in depth, is much better than the control of the vertical when drilling the well. Unfortunately, this exploration technique cannot operate when caves and conduits are not penetrable.

Since the last 30 years many additional techniques for exploring a karst aquifer from the analysis of its behavior are in use in order to overcome these difficulties. Most of them were previously developed for and carried out in surface hydrology. They are very well suited to karst hydrology, because karst aquifers present hydraulic characters similar to those of surface stream systems with separated transmissive and storage functions. These techniques belong to two categories :

- natural functioning studies (hydraulic techniques, natural tracing techniques),
- artificial or induced functioning studies (pumping tests, tracing tests).

Among the complete set of techniques used to study karst aquifers, only methods analyzing natural tracing will be described below

4. NATURAL TRACING TECHNIQUES

4.1. The spring chemograph approach

In comparison to the spring hydrograph, which is the time variation of flow rate, observed chemical variations were named the "chemograph" [Ashton, 1966]. The chemograph does not correspond to the same phenomenon as that producing the hydrograph. The hydrograph is the result of the transfer of the flood wave (the impulse) through the aquifer, produced by rain and infiltration, whereas the chemograph is related to the volume of recharge water flowing through the aquifer (mass flux). The chemograph is induced by the hydrograph, but it does not bear information about the same processes, and it frequently leads to non-significant correlations between total dissolved solid (TDS) and flow rate, contrary to the surface flow [Meybeck, 1984]

Many scientists have tried to identify the causes of chemical variations at karstic springs in order to obtain information about the functioning and flow path organization of studied aquifers. The chemograph is interpreted as the result of the output of different types of qualitatively and quantitatively differentiated waters. Ashton [1966] followed by Andrieux [1972], Bakalowicz and Aminot [1974] and Eberentz [1975] interpreted temporal variations in chemical or isotope characteristics and in temperature as a "flush flow effect" during flood events studied in different karst systems. Bakalowicz and Aminot [1974] and Hess and White [1974] noticed that the shape of the chemograph for a given system is related to the position of the flood within the hydrological cycle. The first flood event of the hydrological year, following the low stage shows the broadest range of chemical variations and hence seems to offer the most interesting information. Chemographs are different eventhough flood events always appear to be chemically well structured. Figure 2 represents the schematic chemical responses described in the literature.

"Model 1" corresponds to the discharge, during a flood event, of water less mineralized than before the flood. Minimum discharge occurs around or after the flood peak. This model applies to surface rivers.

"Model 2" corresponds to the discharge, during flood rising, of water more mineralized than before the flood. Maximum mineralization may occur before or after the flood peak (the doted line on fig.2 corresponds to a maximum mineralisation in phase with the flow peak), sometimes followed by minimum or decreasing discharge, and sometimes disturbed by rapid variations.

"Model 3" shows invariant or very regular chemical variations.

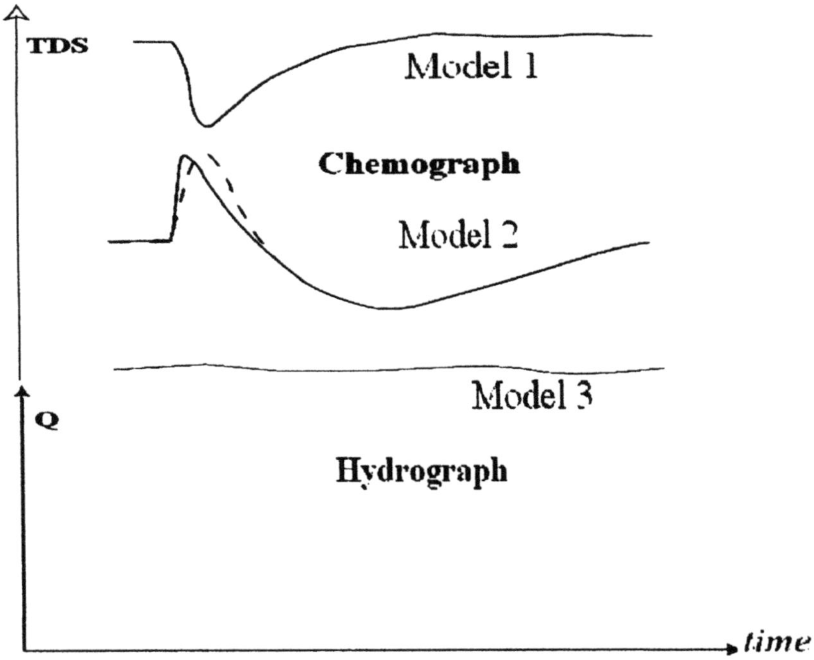

Figure 2. The different types of chemograph, corresponding to different chemical models.

Chemographs and spring hydrographs at the karst outlet seem to be representative of the related karst system. Therefore their analysis should provide a powerful tool for deciphering information about system structure and functioning. Nevertheless, at the present stage of knowledge, specific interpretations of chemographs are given for different karsts studied. In the literature, many interpretations are proposed depending on the karst system studied.

Beyond the specificity of each karst (geological setting, development of karst features, climate, etc.), one may wonder whether there are some common characters and processes involved in the functioning of karstic aquifers which are expressed in spring chemographs? Or, are chemical data sufficient to help determining a general karst functioning scheme?

At the present time, the literature does not clearly identify the process(es) responsible for water chemical changes at a karst spring during a flood. Some processes put forward by different authors include:

1. a flush flow effect concerning the whole aquifer in its infiltration and phreatic zones, i.e. a piston flow effect,
2. a flush flow effect occuring only in the conduit network, and acting as a bypass or shortcut,

3. a mixing effect between the water stored in the phreatic zone and the recharge water,
4. a chemical effect such as CO_2 transport, inducing fast rock dissolution.

The proposed processes concern different types of water with different "histories", resulting in different geochemical compositions, depending on the type of chemical response and ions involved. Then, depending on their extent and location in the karst system, each of these waters may be pushed or flushed away through conduits and cracks or only through conduits, according to the interpretation.

It was found that during flood events karst aquifers generally present (at their outlet) much greater variations in dissolved ions content than porous or fracture aquifers. Natural tracing techniques in karst aquifers were thus introduced by Ashton [1966, 1969], Bakalowicz et al. [1970], Shuster and White [1971], and later by Back and Zötl [1975], Atkinson [1977], Mudry [1981] and Drake [1983]. Since natural tracers provide information about water origin, transit time and/or human impacts, the hydrogeochemical approach to karst systems is increasingly taken into account in studies of karst aquifer functioning and human impacts on karst groundwater quality.

Natural tracers may characterize processes in the infiltration zone, specifically in the epikarst, as well as in the phreatic zone [Plagnes, 1997]. Several geochemical tracers may be studied at the same time. They generally show wide variations during floods at karst springs.

Various processes cause variations in dissolved chemical contents in karst spring water: dilution and mixing are considered key factors, but piston flow, dispersion or solution processes and kinetics are also suspected by some authors [see Plagnes, 1997, for an exhaustive literature review]. No theoretical approach to natural chemical tracing based on models is presently available to analyze and characterize chemical variations. Therefore the technique is used only as a descriptor of system functioning for the purposes of building a conceptual model of this functioning, through empirical interpretations [Bakalowicz, 1994].

Dissolved inorganic ions, which are the most common informative tracers considered in karst hydrology, have their content determined by processes related to four sources: the geological environment (rock solubility, impurities), biosphere and climate (CO_2 production, water temperature, aerosols), human activities (pollution) and flow conditions (flow velocity, one- or two-phase flow, water residence time, water-to-rock contact surface). Natural tracing provides information about water residence time in karst systems, flow conditions and source of dissolved solids.

In carbonate aquifers, parameters related to carbonate equilibria are especially interesting because they reflect groundwater flow conditions. These parameters are the CO_2 partial pressure (pCO_2), corresponding to the dissolved CO_2 content, and the calcite saturation index (SIc), which indicates whether water is undersaturated or oversaturated or at equilibrium with respect to calcite. These two parameters define dissolved Ca and HCO_3 concentrations and pH. The CO_2 partial pressure depends on the initial soil pCO_2, when water infiltrates, or on pCO_2 of air in fissures of the infiltration zone. SIc < 0 means that water residence time in carbonate is very short (several days to a few weeks), and accordingly that the subsurface flow is very rapid. SIc >0 means that outgasing has occurred during subsurface flow, i.e. that groundwater flow was subject to conditions such as air-pCO_2 lower than water-pCO_2 in the infiltration zone, or with respect to free surface flow in karst conduits.

Environmental isotopes are also considered in natural tracing techniques. The complexity of their variation in karst springs requires abundant sampling, which is consequently expensive but may reveal information that cannot be obtained from dissolved ions. Oxygen-18 content, which is a function of the altitude of rainfall and recharge water, informs about the average altitude of the recharge area [Bakalowicz et al., 1974; Mudry, 1981]. This assessed altitude has to be compared with that derived from the determination of system limits using other different techniques. Tritium may be useful to detect the presence of long residence time water [Bakalowicz and Olive, 1970; Payne et al., 1978]. Oxygen-18 and carbon-13 are sometimes considered when deep CO_2 and thermal waters may contribute to karst groundwater flow [Bakalowicz et al., 1989]. The strontium isotope ratio $^{87}Sr/^{86}Sr$ may be useful to characterize a surface flow contribution from a granite massif, which has a different ratio compared to that of marine sediments such as carbonate because of the ^{87}Sr enrichment from ^{87}Rb [Plagnes et al., 1996).

Radon (Rn) is a natural tracer of karst aquifer which has been recently introduced. Three different radioactive isotopes of radon are present in the Nature. However ^{222}Rn, with its 3. 82 day half-life is the only isotope that is commonly used as a tracer. ^{222}Rn is the gaseous daughter of ^{226}Ra which itself is a daughter product of the decay-chain of ^{238}U. Accordingly, since U is an omnipresent element, so is Rn. In addition, Rn is the only natural radioactive element which is an alpha emitter. As a consequence, it can be detected and measured in the ground or in the groundwater without interference with any other radioactive element. Its concentration or its flux can be measured either in a dicreet way or continuously using in situ electronic automatic devices [Monnin and Seidel, 1998]. Another useful property of Rn is the fact that it is a noble gas which, accordingly, does not interact chemically with the bedrock or with the dissolved elements. Therefore it reflects the water flow intensity and/or the origin of waters. Most of the radon present in karst water is from surface origin or from clay filling fractures and karst cavities. Sometimes, Rn can be carried up by deep origin water and the variation of its concentration is a good indication of mixing. Several work regarding the use of radon as a natural tracer of aquifers have been published [see e.g. Andrews et al. ,1991; Eisenlohr and Surbeck ,1995; Hoehn and von Gunten, 1989; Morin, 1992; O'Connor et al., 1993; Pane ,1995; Pane et al. ,1994; Pane-Escribe, 1995]. A typical example of rainfall related Rn concentration in the outlet water of a karst system is displayed in Figure 3. It can be seen that every rainfall event is followed within a few days by a steep increase in the Rn content (up to 8 fold the initial content). It can, also be seen that the subsequent decrease of the Rn content is not as dramatic and follows a smooth curve along periods of time that

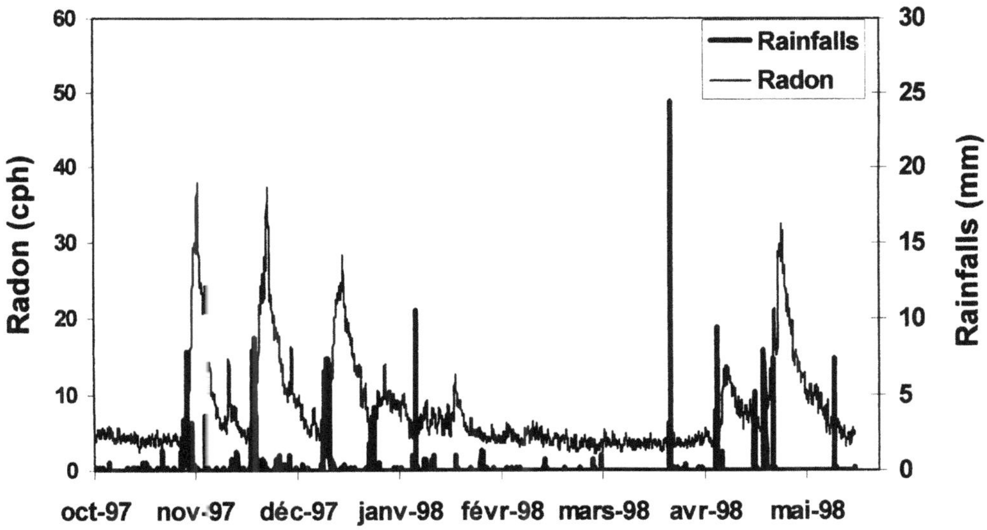

Figure 3. Radon concentration related to rainfall.

extent to several weeks. The first part of the response curve is related to the main transmissivity of the system, while the second part is related to its storage capability.

Dissolved gases are also present in karst groundwater. The distribution of the atmospheric gases and carbon dioxide in the waters can provide useful informations on:

1. the continental water circulation in karst systems;
2. the influence of other parameters such as energy, chemicals (whether in solutions or as suspended matter), sediments, pollutants, etc;
3. the response of such hydrosystems to external (natural or anthropic) stresses.

The main atmospheric constituents are nitrogen (N_2 = 0.78084 volume fraction of dried air), oxygen (O_2 = 0.20946), argon (Ar = 9.34×10^{-3}) and carbon dioxide (CO_2 = 3.5×10^{-4}). Moreover noble gases are also present but in lower concentrations: neon (Ne = 1.8×10^{-3}), helium (He= 5.24x10-6), krypton (Kr = 1.13×10^{-6}), xenon (Xe = 8.7×10^{-8}) [Charlson, 1992].

The amount of gas which dissolves in water is a function of the ambient temperature and pressure, the latter being a function of altitude. In addition, once groundwater enters the saturated zone its atmospheric noble gas content remains constant.

Concentration of Ar,O_2, N_2, and CO_2 can be determined by gas chromatography. ^{4}He, ^{20}Ne, ^{22}Ne, ^{36}Ar, ^{38}Ar, ^{40}Ar can be determined on a quadrupole mass spectrometer after separation of noble gases from chemically reactive gases : H_2S, CO_2, N_2, H_2, SO_2, CH_4, etc.

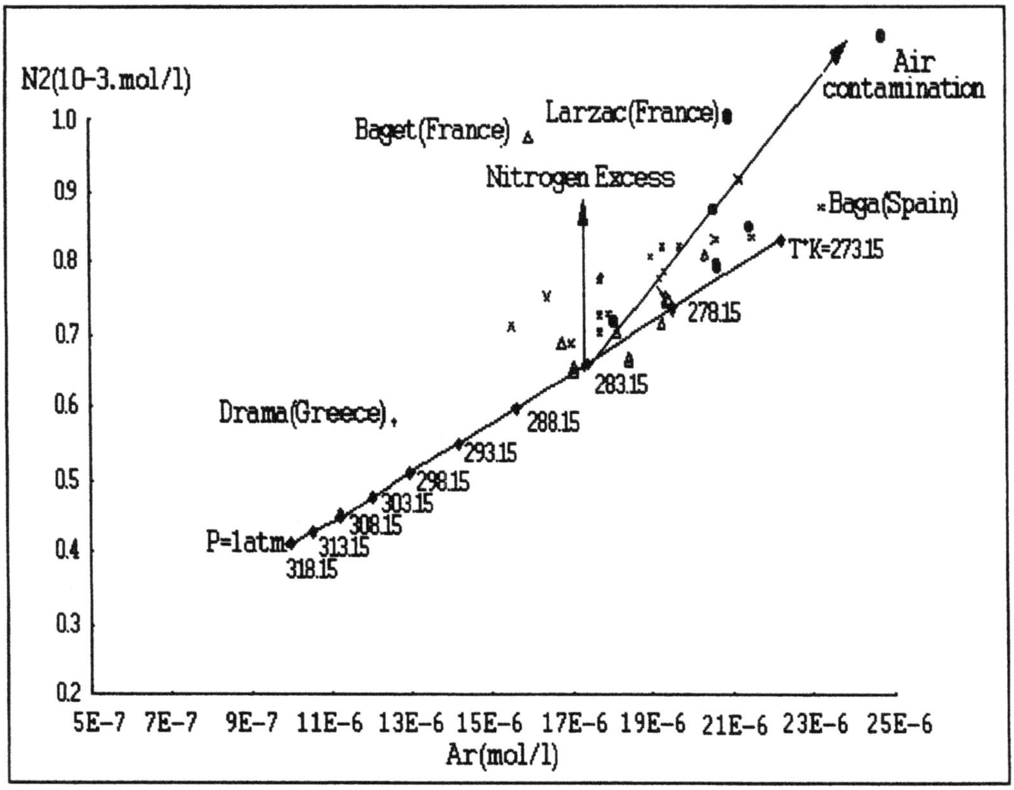

Figure 4. Argon plot from Nitrogen. The line corresponds to the equilibrium ratio from 0 to 45°C at 1 atmposhere.

The nitrogen versus argon diagram shown in Figure 4 was obtained by the IGGI of Pisa (Italy) within the frame of a European Research Programme on Karst [Monnin, 1998]. The measurements were carried out on several karst systems. Most samples plot over the line that expresses the N_2/Ar molar ratios in the gas phase in equilibrium with saturated pure water from 0 to 45°C and 1 bar. This pattern can be due to two different processes: (i) contamination with air and/or (ii) input of nitrogen produced by de-nitrification processes in soils. Even if contamination with air can occur either at the moment of sampling and during sample transportation, storage and analysis, a different mechanism has been proposed to explain the excess of air dissolved in water, commonly encountered in karst systems. According to Herzberg and Mazor [1979], "in karstic systems air occurs in fractures voids of the rocks, along with down-flowing water. When the water flows through narrow channels the air in the system is sucked as excess air. In the high-discharge season most of the system is filled with water and less air gets sucked in. The situation is reversed in the low-discharge season when more space for air is available. Hence the negative correlation of Neon excess to discharge is obtained. The effect is most obvious for the least soluble gas, Ne".

Other natural tracers are also considered in the literature. The "drift of microcrustaceans" at karst springs was analyzed [Rouch, 1980; Gibert, 1986; Gibert et al., 1988; Moeschler et al., 1982] not only for studying the population dynamics, but also as tracers of water origin. Some of these microanimals may inform about recharge by surface streams through swallow holes, or about water from conduit or storage zones [Rouch et al., 1993]. Inorganic suspended particles may also inform about the source or functioning. However, processes responsible for transport of suspended particles are much more complicated to understand and difficult to measure.

4.2. The main natural tracers in karst spring chemographs and their meaning

Instead of referring only to a general tracer of chemical variations such as electrical conductivity lower (EC) or total dissolved content (TDS), it is also interesting to refer to specific tracers, as briefly shown in some of the cited examples. According to the considered tracer and to the karst system, inorganic dissolved ions and environmental isotopes carry information about chemical processes related to flow conditions and sources. In order to guide karst investigators, we propose — based on our experience — a rough assessment of the most common tracers, as summarized in Table 1. Each of them may characterize one or several processes according to the type of variation.

4.3. Geochemical interpretative methods

All authors attribute chemical variations observed at springs to the role of fast infiltration and/or to the conduit network, which acts as a preferential pathway for surface and rain water. Comparison of chemical variations, expressed by the total dissolved solid content (TDS) in water, to the corresponding hydrograph shows a highly significant relationship between chemical and flow rate variations (Fig. 2).

Shuster and White [1971], followed by Ternan [1972], proposed a variation index of total hardness or TDS to characterize carbonate aquifers, as conduit flow or diffuse flow systems. However, the comparison of variation index values is statistically significant only if the distributions to which the data pertain are defined by the same law. Unfortunately, as shown by Bakalowicz [1977], the distribution laws are different and may even be multimodal. Consequently, the variation index cannot be used to characterize carbonate aquifer behavior.

Using a geostatistical approach, monitoring of spring water chemical composition at two different time scales, i.e. long-term monitoring over one or several hydrological cycles (weekly or monthly sampling) or short-term monitoring during flood events (hourly or daily sampling), should give the same information related to the system functioning and consequently about karst conduit extension, inside the aquifer. The technique is used as a descriptor of system functioning to build a conceptual model of this functioning by means of empirical interpretations [Bakalowicz, 1994]. The width and multimodality of the distribution - of which the most common related parameter is the electrical conductivity (EC) - depend on the

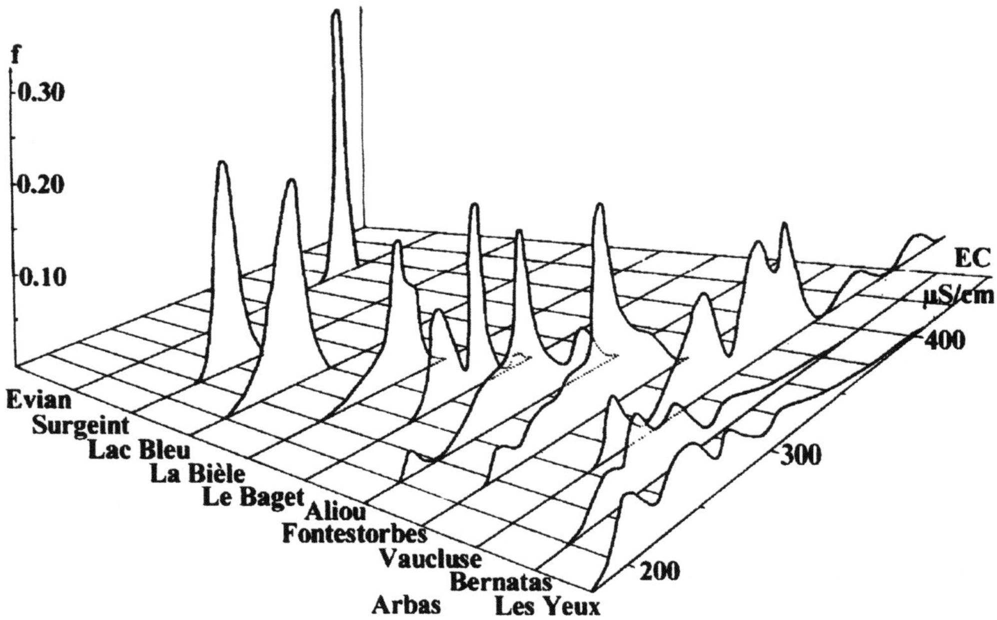

Figure 5. Distributions of mineralization at karst springs.

development of the functional karst structure. Therefore karst system functioning may be characterized by classifying these systems according to karstic network development (Fig. 5). A porous carbonate aquifer (e.g. Evian) is characterized by a very narrow unimodal distribution. A fracture carbonate aquifer is characterized by a narrow unimodal distribution. In contrast, a karst aquifer is characterized by a wide multimodal distribution with width and number of modes increasing together with karst conduit development.

Statistical methods, such as multidimensional analysis (principal component analysis (PCA), discriminant factor analysis (DFA), and others may be used to describe and analyse chemical data series from karst springs [Bakalowicz, 1984; Mudry and Blavoux, 1986; Mudry, 1990] with the aim of building conceptual models.

Artificial tracing tests are utilized in karst since long ago. They were firstly used, and still are sometimes, to show the existence of a hydraulic relationship between an input (a swallow hole) or an underground stream and a spring. Salt and fluoresceine were the most common tracers. Transport of artificial tracers in karst was early theorized, in order to obtain more information from the experiments than from the qualitative tests. This was possible, because accurate analysis of tracers at very low contents could be carried out efficiently, thanks to new chemistry technologies. Breakthrough curves are the basic data, from which balance and recovery rate, first

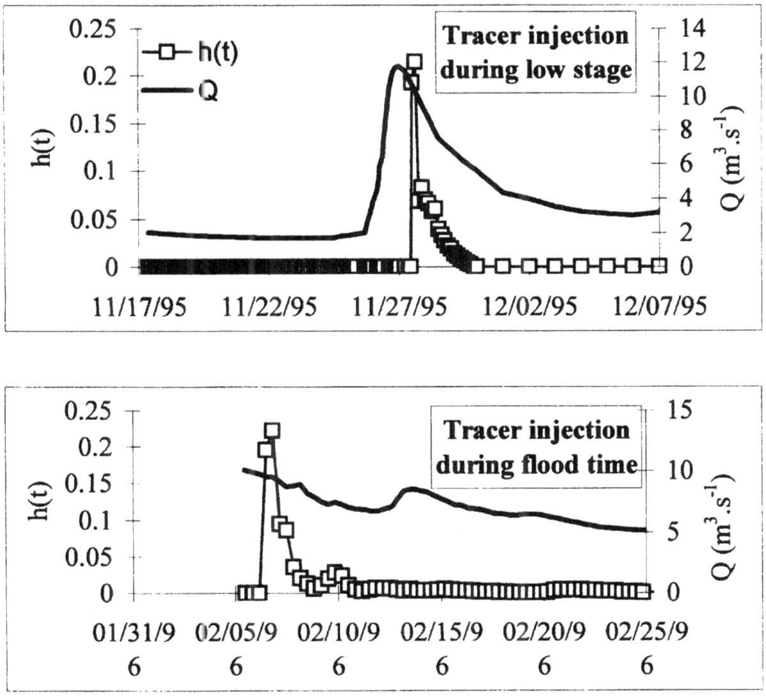

Figure 6. Example of breakthrough curves of tracing tests. Tracer was injected in the same swallow hole.

apparition, mode and mean times, and their corresponding velocities are calculated [Käss, 1998]. Theoretical approaches of breakthrough curves proposed analytical methods for their interpretation, in terms of dispersion and diffusion in modeled media [Zuber et al., 1995]. These approaches were abundantly discussed [Atkinson et al., 1973, 1979 ; Meus, 1993 ; Meus and Bakalowicz, 1997 ; Kass, 1998]. As a consequence of karst aquifer complexity, it seems that the present best approach is an empirical analysis of the breakthrough curve (Fig. 6), as proposed by Meus [1993] and Field [1999]. The method developed by Field [1989] for the American Environmental Protection Agency (EPA) must be recommended for karst studies ; it proposes a free programme package (QTRACER) which gives the most important parameters, with a minimum of basic realistic assumptions.

5. CONCLUSION : IS THIS METHODOLOGY EFFICIENT FOR KARST EXPLORATION?

Finally, all the sets of observations and data obtained from the different techniques put in operation for exploring the karst system must be linked together by the karst hydrogeologist in order to build a conceptual model of the system. It is the phase of diagnosis, well known from physicians. The conceptual model gathers together and synthesizes all the investigations pieces of informations.

However, most of the presented techniques are not yet completely based on theoretical approaches suitable to karst characteristics. Up to now, the proposed methodology for the exploration of its groundwater resources follows an almost empirical way. Some of the techniques show problems in their application, which are briefly pointed out, in order to indicate some research directions.

Chemograph seems to be an interesting descriptor of flow and transport through karst aquifer during flood. Unfortunately, if the main involved processes are identified, no analytical method interpret it thoroughly.

Methods for interpreting tracing tests must obviously be adapted to the karst specificity. They cannot be based on classical models whose basic assumptions are never met in karstic medium. Nowadays no method proposes an analytical approach to characterize the traced or pumped part of the system.

Karst systems present extensions up to several hundreds km^2, much wider than other aquifers, onto which any human activity may have impacts within a short time (few weeks to several months) and at long distance. This is the reason why the exploitation of karst groundwater resources requires to consider large regions when implementing managing plans. Geographic information systems which link the data together and to maps are an efficient and necessary tool as an assistant for decision makers.

In classical hydrogeology, modeling is the final tool which allows the simulation of the aquifer functioning, in its whole or in its parts, under the effect of changes (recharge, discharge, pollution, etc.). By now, despite important scientific efforts, modeling of karst aquifers functioning remains in the research domain and cannot be considered as an exploration technique, even if a few examples can be cited [Palmer et al., 1999].

Monitoring systems and sampling procedure are preferred to simulation by complicated models, whose boundary conditions do not fit well to karst aquifers. They allow warning when a change occurs at the system scale, because most of the time karst reacts very fast. Such fast reactions of karst aquifers are considered as a major negative character : karst is commonly declared as very vulnerable to pollution or very sensitive to water withdrawals. But, by comparison with porous and fracture aquifers, which dynamics is slow, karst shows a generally weak resilience, *i.e.* it may come back rapidly to its initial conditions after stopping the disturbing effects. These changes have to be monitored in order to fit the arrangements in the best way.

At last, karst hydrogeologists have to set up and implement a complex set of techniques, for describing karst system extension and limits, exploring its drainage pattern, and analyzing its behavior. All geoscience disciplines are thereby required for the exploration of groundwater resources in karst aquifers. Technical tools exist for such anexploration. But progress is still necessary in the theoretical approach of karst functioning and in karst modeling in order to improve the interpreting methods for analyzing data and observations in a more efficient way.

REFERENCES

Andrieux, C. (1972) Recherches sur les eaux souterraines. 18 - Le système karstique du Baget (Ariège). I. Sur la thermique des eaux au niveau de l'exutoire principal (note préliminaire). *Annales de Spéléologie*, 27(3): 525-541.

Andrews, J.N., Drimmie, R.J., Loosli, H.H. and Hendry, M.J. (1991) Dissolved gases in the Milk River aquifer, Alberta, Canada. *Appl. Geochem.*, **6**, , 393-403.

Ashton, K. (1966) The analysis of flow data from karst drainage systems. *Transactions of Cave Research Group Great Britain*, 7: 163-203.

Ashton K. (1969) The present position of the theory and technique of pulse wave hydrology. 5th Internat. Congress of Speleology, 1-3.

Atkinson, T.C. (1977) Diffuse flow and conduit flow in limestone terrain in the Mendip Hills, Somerset, Great Britain. *J. Hydrol.* 35, 93-110.

Atkinson T. C., Smith D. I., Lavis J. J., and Whitaker R. J. (1973) Experiments in tracing underground waters in limestones. *J. Hydrol.* 19, 323-349.

Atkinson T. C. and Smart P. L. (1979) Traceurs artificiels en hydrogéologie. *Bulletin B.R.G.M., 2, section III* **3**, 365-380.

Back W. and Zötl J. (1975) Application of geochemical principles, isotopic methodology and artificial tracers to karst hydrology. In *Hydrogeology of karstic terrains*, Vol. B, 3 (ed. A. B. L. Dubertret), pp. 105-121. IUGS.

Bakalowicz M. (1977) Etude du degré d'organisation des écoulements souterrains dans les aquifères carbonatés par une méthode hydrogéochimique nouvelle. *C. R. Acad. Sci. Paris, D* **284**, 2463-2466.

Bakalowicz M. (1984) Water chemistry of some karst environments in Norway. *Norsk geografisk Tidsskrift* **3**(4), 209-214.

Bakalowicz M. (1994) Water geochemistry : water quality and dynamics. In *Groundwater ecology*, Vol. 1 (ed. J. Stanford, J. Gibert, and D. Danielopol), pp. 97-127. Academic Press.

Bakalowicz M. (1996) Les processus de karstification et les différents types de karst associés. *Mém. Soc. Géol. France*(169), 363-371.

Bakalowicz M. (1997) Hydrogéologue versus spéléologue, ou de qui relève l'étude et l'exploration des eaux souterraines karstiques? 6th Conference on limestone hydrology and fissured media, 23-26.

Bakalowicz M. (1999) Connaissance et gestion des ressources en eaux souterraines dans les régions karstiques. Agence de l'Eau Rhône - Méditerranée - Corse.

Bakalowicz M. and Olive P. (1970) Teneurs en tritium des eaux du karst du Taurus occidental et de Pisidie (Turquie). *Schweizerische Zeitschrift für Hydrologie* **32**(2), 475-480.

Bakalowicz, M. and Aminot, A. (1974.) Géochimie des aquifères karstiques. 3. Premiers résultats obtenus sur le système du Baget durant le cycle hydrologique 1973. *Annales de Spéléologie*, 29, 4 : 484-493.

Bakalowicz M., Ford D. C., Miller T., Palmer A. N., and Palmer M. V. (1989) Thermal genesis of dissolution caves in the Black Hills, South Dakota. *Geol. Soc. Am. Bull.* **99**, 729-738.

Benderitter Y. (1997) Karst et investigations géophysiques. *Hydrogéologie* (3), 19-30.

Biswas A. K. (1970) History of hydrology. Amsterdam, London: North Holland.

Charlson R.J. (1992) The atmosphere. *In* « Global biogeochemical cycles ». Butcher S.S., Charlson R.J., Orians G.H. and Wolfe G.V. Eds. pp. 213-238. Academic Press, Harcourt Brace Jovanovich, London.

Crochet P. and Marsaud B. (1997) Approches conceptuelles de l'aquifère karstique. Problèmes méthodologiques et d'exploitation. *Hydrogéologie* (3), 3-18.

Cvijic J. (1893) Das karstphaenomenon. Versuch einer morphologischen Monographie. *Geog. Abhandl.* Wien 5 (3), 218-329.

del Zanna, P. (1899) I fenomini carsici nel bacino dell'Elsa. *Boll. Soc. Geol. Italiana,* XVIII, 1-11.

de Lapparent, A. (1883) Traité de géologie. Première partie : phénomènes actuels. Masson, Paris, 3ème édition.

Drake J. J. (1983) The effects of geomorphology and seasonality on the chemistry of carbonate groundwater. *J. Hydrol.* (61), 223-236.

Eberentz, P. (1975) Apport des méthodes isotopiques à la connaissance de l'aquifère karstique, Univ. Paris VI: 47.

Eisenlohr, L. and Surbeck, H. (1995) Radon as a natural tracer to study transport processes in a karst system. An example in the Swiss Jura. *C.R. Acad. Sci. Paris,* **t.321**, série II a, 761-767.

Field M. (1999) The QTRACER program for tracer-breakthrough curve analysis for karst and fractured-rock aquifer. Nat. Center for Environ. Assess., U.S. EPA.

Ford D. C. and Williams P. W. (1989) *Karst geomorphology and hydrology.* Acad. Div. Unwin Hyman Ltd.

Gibert J., (1986) Ecologie d'un système karstique jurassien. Hydrogéologie, dérive animale, transits de matières, dynamique de la population de Niphargus (Crustacé Amphipode). Doctorat ès Sciences naturelles, Univ. Claude Bernard, Lyon.

Gibert J., Mangin A., and Bakalowicz M. (1988) A system analysis method for ecological studies of karsts. Example : Dorvan-Cleyzieu karst (Southern Jura, France). *Oikos* **53**(2), 1-18.

Grund A. (1903) Die Karsthydrographie: Studien aus Westbonien. Geog. Abh. Heraus. Von A Penck. E, 103-200.

Hess J.W. and White W.B. (1974) Hydrograph analysis of carbonate aquifers. Inst. for Res. on Land and Water Resour., the Pennsylvania State University, Pennsylvania.

Herzberg O. and Mazor E. (1979) Hydrological Applications of noble gases and temperature measurements in underground water systems: examples from Israel. *J. Hydrol.* **41**, 217-231.

Hoehn, E. and von Gunten, H.R. (1989) Radon in groundwater : a tool to assess infiltration from surface waters to aquifers. *Water Resour. Res.*, **25**, 8, 1795-1803.

Hutton J. (1795) Theory of the Earth, with proofs and illustrations, vol II, Edinburgh.

Jakucs L. (1977) *Morphogenetics of karst region : variants of karst evolution.* Akademiai Kiado.

Kaas W. (1998) *Tracing technique in geohydrology.* Balkema.

Kruber A. (1900) Phénomènes du Karst en Russie. *Semlewjedjenie,* Moscou VII, 4, 1-34.

Mangin, A. (1974) Notion de système karstique. *Spélunca Mémoires,* 8, 65-68.

Mangin A. (1975) Contribution à l'étude hydrodynamique des aquifères karstiques. Doctorat ès Sciences naturelles, Université de Dijon, France.

Mangin, A. (1978) Le karst, entité physique, abordé par l'étude du système karstique. Le karst, colloque de Tarbes, 17-18 octobre 1978, AGSO, p.21-37.

Mangin A. (1982) Pour une meilleure connaissance des systèmes hydrologiques à partir des analyses corrélatoires et spectrales. *J. Hydrol.* **67**, 25-43.

Meus P. (1993) Hydrogéologie d'un aquifère karstique dans les calcaires carbonifères (Néblon-Anthisnes, Belgique). Apports des traçages à la connaissance des milieux fissurés et karstiques. Doctorat en Sciences, Université de Liège.

Meus P. and Bakalowicz M. (1997) Les traçages artificiels, outils de reconnaissance et d'étude des aquifères karstiques. *Hydrogéologie* (3), 43-50.

Meybeck M. (1984) Les fleuves et le cycle géochimique des éléments. PhD Thesis, Univ. Paris-6, 477 pp.

Moeschler P., Müller I., Schötterer U., and Siegenthaler U. (1982) Les organismes vivants, indicateurs naturels dans l'hydrodynamique du karst, confrontés aux données isotopiques, chimiques et bactériologiques. *Beitrage zur Geologie der Schweiz, Hydrogeologie* **23**, 213.

Monnin M. (1998) A multidisciplinary global approach of groundwater flows in karstic areas and its consequencers for water resources and environmental studies. Rept. ERB-CHRX-CT94-0567 European Commission DGXII - Brussels.

Monnin M. and Seidel J.L. (1998) An automatic radon probe for earth science studies. J. *Appl. Geophys.* 39, 209-220.

Morin J.P. (1998) Le radon comme traceur du transport des fluides dans le sol : simulation, modélisation et applications géophysiques. PhD Thesis, Clermont II.

Mudry J. (1981) Sur l'origine des gradients des teneurs isotopiques et géochimiques dans les eaux karstiques du Jura (France). *J. Hydrol.* **50**, 167-178.

Mudry J. and Blavoux B. (1986) Utilisation de l'analyse en composantes principales (sur variales centrées réduites) pour l'étude du fonctionnement hydrocinématique de trois aquifères karstiques du Sud-Est de la France. *Hydrogéologie* **1**, 53-59.

Mudry J. (1990) Les courbes flux chimique-débit et le fonctionnement des aquifères karstiques. *J. Hydrol.* **120**(1-4), 283-294.

Nicod J. (1972) Pays et paysages du calcaire. P.U.F. - Paris

O'Connor, P.J. et al. (1993) Assessment of the geological factors influencing the occurence of radon hazard areas in a karstic region. 93-2, Geological Survey Ireland, Dublin.

Palmer A. N., Palmer M. V., and Sasowsky I. D. (1999) Karst modeling, pp. 265. Karst Water Institute.

Pane M. B.(1995) Le radon traceur des circulations en milieu karstique. *C.R. Acad. Sci. Paris*, **340**, 4, 37-45.

Pane M.B., Seidel J.L., Monnin M. and Morin J.P. (1994) Radon as a tracer of fluids motion in fractured aquifers. *Environ. Geochem. & Health*, **16**, suppl., 325-334.

Pane M.B. (1995) Utilisation des éléments en trace comme traceurs des circulations souterraines en milieu karstique (site du Lamalou, Hérault). PhD thesis, Montpellier 2, Montpellier, 297 pp.

Payne B. R., Leontiadis J., Dimitroulas C., Dounas A., Kallergis G., and Morfis A. (1978) A study of the Kalamos springs in Greece with environmental isotopes. *Water Resour. Res.* **14**(4), 653-658.

Penck A. (1900) Geomorphologische Studien aus der Hercegovina, Z. *Deut. Osterreich. Alpenver* 31, 25-41.

Penck A. (1903) Über das Karstphänomen. *Verbreit. Naturwis. Kenntniss*, Vienne, 1.

Pfeiffer D. (1963) Die geschichtliche Entwicklung des Anschgauungen unber das karstgrund-wasser. Beihefte zum geologischen Jahrbuch 57 (Hannover).

Plagnes V. (1997) Structure et fonctionnement des aquifères karstiques. PhD thesis, Université Montpellier II.

Renault P. (1967) Contribution à l'étude des actions mécaniques et sédimentologiques dans la spéléogenèse. I - Introduction. Le problème de la spéléogenèse. *Annales de spéléologie*, **22**(1), 5-17.

Rouch R. (1980) Les Harpacticides, indicateurs naturels de l'aquifère karstique. *Mémoires hors série de la Société géologique de France* **11**, 109-116.

Rouch R., Pitzalis A., and Descouens A. (1993) Effets d'un pompage à gros débit sur le peuplement d'un aquifère karstique. *Annal. Limnol.*, **29**(1), 15-29.

Shuster E. T. and White W. B. (1971) Seasonal fluctuations in the chemistry of limestone springs: a possible means for characterizing carbonate aquifers. *J. Hydrol.*, **14**, 93-128.

Sweeting M. M. (1973) Karst landforms. Columbia University Press.

Ternan, J.L. (1972) Comments on the use of a calcium hardness variability index in the study of carbonate aquifers: with reference to the Central Pennines, England. . *J. Hydrol.*, **16**, 317-321.

von Mojsisovics (1880) Die Karst-Erscheinungen. *Zeitschrift des deutschen Alpenvereins*,.

Zuber A., Motyka J., Osenbrück K., Weise S. M., Grabczak J., and Mikolajczyk H. (1972) Regional transport of solutes in fissured rocks : a reconnaissance study of a limestone aquifer in Cracow, Poland. *Archiwum Ochrony Srodowiska* **1**, 61-72.

Land and Marine Hydrogeology
Edited by M. Taniguchi, K. Wang and T. Gamo

Abundance and Viability of Subsurface Microbial Communities in Sedimentary and Igneous Rock Aquifers

Y. Murakami[a], Y. Fujita[a], T. Iwatsuki[b], and T. Naganuma[a]

[a]School of Biosphere Sciences, Hiroshima University, Higashi-hiroshima, 739-8528 Japan

[b]Tono Geosciene Center, Japan Nuclear Cycle Development Institute, Toki, 509-5102 Japan

Abundance and viability of microorganisms in groundwater were studied using a geochemically characterized borehole in the Tono uranium deposit, central Japan. Using autoclaved samplers, groundwater samples were collected from the scientific borehole TH-6 at 4 depths: 104, 132 and 153 m in sedimentary rocks and 177 m in granite rock. Total cell counts as determined by epifluorescence microscopy were of the order of 10^5 to 10^6 cells ml^{-1}, showing no consistent increase or decrease with depth. Total cell counts were also determined by using 4 different fluorescent dyes: acriflavine, acridine orange, DAPI and SYBR Green I, in three different years. The year-to-year variation in total cell counts was large at the shallow sampling depth. Although there was generally a good correlation between total counts assured from the different dyes, there were sometimes significant differences between total counts measured for each of them. These differences suggest that: total counts may be subject to temporal changes due to hydrological and geochemical changes of groundwaters, which in turn could also affect the microbial composition and/or their affinity to specific dyes. Viability based on cell membrane integrity was 32.4-59.2%, showing a consistent increase with depth. In contrast, viability determined by the activities of electron transport system (ETS) and esterase showed inconsistent depth-profiles. The ETS-active cell counts were 1.8-21.5% of total counts, with the maxima at the uppermost and lowermost sampling depths; the minimum was at 153 m deep, above the sediment-granite unconformity. The esterase-active cell counts were 0.4-9.5% of the total, also showed a minimum at 153 m deep; but the maximum was found at 132 m deep, within the organic-rich Toki lignite layer. Viability based on cultivation was assessed by using the Biological Activity Reaction Tests (BART™) and by the most probable numbers (MPNs). BART and MPN estimates showed different depth-profiles, particularly for sulfate-reducing bacteria (SRB). These differences suggest that there were different community structures of microorganisms at each depth. The MPN-estimates of aerobic and anaerobic sulfur-oxidizing bacteria also showed different depth profiles, suggesting a difference in electron acceptors for sulfur oxidation and that depth-specific microflorae may develop in response to availability of these acceptors (and donors) in the subsurface.

1. Introduction

Abundance, diversity and activity of subsurface microbial communities have been a current focus in geomicrobiology. Biomass of total subsurface microorganisms has been estimated to be 2.0-5.18 x 10^{17} g (Gold, 1992; Whitman et al., 1998), which closely compares with the biomass of traditionally studied terrestrial and aquatic plant and prokaryotes (5.9 x 10^{17} g C; Whitman et al., 1998). These estimates were based upon pioneering studies that measured the microbial abundance in terrestrial sedimentary (Hazen et al., 1991) and aquifers of sedimentary rocks and igneous rocks (Pedersen et al., 1990, 1996, Stevens and McKinley 1995).

It is generally accepted that the lower (deeper) limit of the subsurface biosphere may be outlined by the 110-150°C isotherm, of which is likely the upper temperature limit for life (Stteter et al., 1990). Assuming a thermal gradient of 20-30°C km^{-1} in terrestrial environments (Press and Siever, 1986), this isotherm would lie at ~ 4-5 km depth (Garland, 1979). Therefore it is increasingly believed that the subsurface biosphere is thicker than terrestrial and oceanic regimes. In addition to temperature, pore space in the rock and sediment is a constraint for the distribution of subsurface microorganisms (Fredrickson et al., 1997). A porosity of 50% or more is common for rocks in the upper 5 km of the earths crust. For example, a porosity of >60% is presumed for a 5 km-deep sub-seafloor layer (Floyd et al., 2001). These rock pores provide the volumetric basis for housing a great biomass of subsurface microorganisms.

Diversity and activity of subsurface microbial communities have often been studied in relation to biogeochemical processes such as reduction of sulfate, nitrate and Fe(III) (*e.g.*, Pedersen, 1993; Nealson and Saffarini, 1994). Anaerobic metabolisms should be the dominant forms of microbial energy transfer, material cycling and carbon turnover in the subsurface. For example, the green house gases such as carbon dioxide and methane are released and assimilated by anaerobic subsurface microorganisms (Pedersen, 1993). Evaluation of these subsurface microbial processes should therefore be conducted in relevance to global biosphere.

In addition, mineral-microbial interactions potentially control subsurface hydrogeology. Precipitation and dissolution of various minerals are known to be mediated by microorganisms (Fredrickson et al., 1998; Lovely, 2000). A member of the genus *Shwanella*, to which common subsurface species belong, reduces and dissolve geothite-Fe(III) (Lower et al., 2001). Some subsurface microbial populations from the Tono area facilitate the formation of Fe (III)-colloids (Naganuma et al., 2000). Certain microorganisms corrode granular F^0-barrier and alter the F^0-barrier's hydraulic characteristics (Liang et al., 2000). These examples indicate that microbial metabolism can affects the hydrogeologic features and processes in the subsurface.

The Tono uranium deposit area was chosen as the site for this study. A series of hydrogeological and geochemical investigations have been made of the Tono area, and therefore provide important environmental and geological background information for geomicrobiological studies. A total of 14 deep boreholes have been cored and maintained for

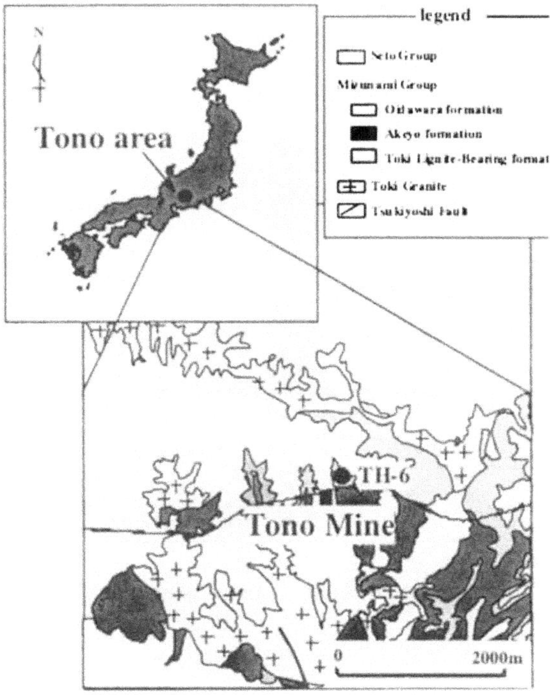

Figure 1. Location of the borehole TH-6 (filled circle) in the Tono uranium deposit area, central Japan. Tono Mine is indicated as a square. TH-6 is hydrogically upgradient from Tono Mine.

monitoring hydrological and geochemical properties of the Tono granite bedrock. The depth of the boreholes ranges from 200 m in the sedimentary rocks to 1000 m in the underlying granitic basement rock. These investigations have been done to assess the impact of subsurface waste disposal on the natural subsurface environment. It should be stated that the Tono area is only a basic study site, not the site of disposal. Although previous investigations have focused mostly on abiotic aspects of the subsurface environment, the demand for subsurface microbial surveys has been increasing (Japan Nuclear Cycle Development Institute, 2000).

This study was conducted within the framework of assessing the subsurface microbial dynamics in the Tono area. This communication reports the results from the first 3-year survey, focusing on the abundance (total counts) and viability (viable counts) of the subsurface microbial communities from a scientific borehole in the area. Temporal and spatial

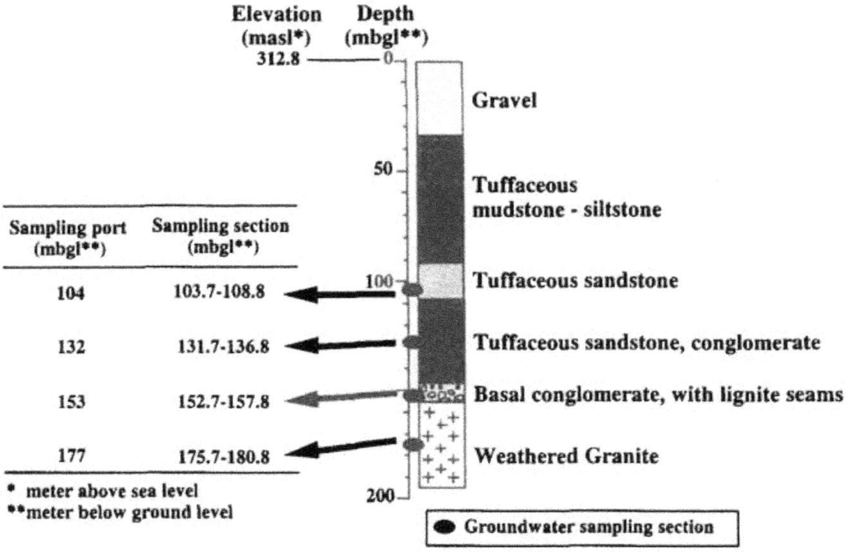

Figure 2. Cross-section of the Tono uranium deposit area, centraland lithologic log of the cored rock samples.

variability of the total and viable microbial counts is discussed.

2. Materials and Methods

2.1. Site description

The Tono uranium deposit area is located in the Toki Basin, central Japan, at about 35.4°N and 137.2°E (Figure 1). The 200 m-deep borehole, TH-6, is slightly north of the Tono Mine (operated by Japan Nuclear Cycle Development Institute, JNC; former Power Reactor and Nuclear Fuel Development Corporation, PNC). TH-6 is the only borehole that is hydrologically upgradient from Tono Mine. Therefore the hydrological disturbance by the mine tunnels is relatively minor when compared with the downgradient boreholes. Also, as noticed in previous studies, TH-6 is the only borehole that yields the smell of hydrogen sulfide. Therefore, we expected to detect activities of sulfate-reducing bacteria in this borehole. A total of 203 m of the borehole was drilled from an altitude of 312 m of the Toki Basin sediment (Figure 2). The geology of Toki Basin is characterized by a thin top-layer of the sand-gravel-dominated Toki Formation of the Seto Group; the lower Seto Group is known to yield porcelain clay for the local pottery industry. Borehole TH-6 penetrates the Tertiary sedimentary rocks of the Seto Group (0.7-5 Ma) and the Mizunami Group (15-20 Ma). These rocks consist of the marine mudstone-dominant Oidawara Formation, marine sand-/mudstone Akeyo Formation, tuffaceous Upper Toki Lignite-bearing Formation, and lacustrine basal conglomerate Lower Toki Lignite-bearing Formation (Fig. 2). The unconformity between the sedimentary and granite rock occurs at about 173m depth, with a 20 m-thick overlying layer of weathered granite. This weathered zone contains a uranium deposit (U_3O_8) estimated to be 4,590 tons. This deposit accounts for 43% of the total Japanese deposit of 7,701 tons of U_3O_8 (Power Reactor and Nuclear Fuel Development Corporation, 1994). The borehole ends in the Toki Granite bedrock (55-65 Ma) with about 30-m penetration.

2.2. Groundwater sampling and geochemistry

The borehole TH-6 was drilled using local freshwater as the drilling fluid, instead of conventional drilling mud fluids, thereby minimizing the mud-derived contamination of the groundwater samples (Iwatsuki and Yoshida, 1999a). Sodium fluorescein (uranine) was added to the drilling fluid to monitor the residual contamination by the drilling fluid, and the drilling fluid was pumped out from the borehole. The borehole was repeatedly 'flushed' by pumping with the in situ formation water before and after casing in order to replace the residual drill fluid. Flushing with 3-well-volumes of formation water should allow the borehole to represent undisturbed groundwaters (Fredrickson and Phelps, 1997). However, after borehole TH-6 was cased (inner diameter, 38 mm; polyvinyl chloride) and packed with the multiple-piezometer monitoring and fluid sampling system (MP System, Westbay Instruments Inc., North Vancouver, Canada) in 1988, an unrecorded volume of groundwater was subsequently pumped out from each of the packer sections.

Using the cased-and-packed borehole TH-6, groundwater samples were collected through the sampling ports at 4 depths of 104, 132, 153 and 177 m below ground level. The 104 m-deep port was located in middle-grained sandstone of the upper Akeyo formation (16-17 Ma). The 132 m-deep port was positioned in the coarse-grained sandstone and conglomerate stratum of the mid-Akeyo Formation (18-20 Ma). The 153 m-deep port targeted the depth of uranium deposit above the sediment-granite rock unconformity at 173 m depth. The 177 m-deep port, beneath the unconformity, was placed in the upper Toki Granite bedrock (60-100 Ma). Hydrogeological characteristics of the aquifers and rocks at the sampling depths are summarized in Table 1. The groundwater in the sedimentary rocks is believed to migrate downward. The time required for the migration from the upper sedimentary rock (Upper Toki lignite-bearing Formation) to the lower layer (Lower Toki lignite-bearing Formation) is estimated to be approximately 10000 years. In addition, part of the groundwater in the lower layer may be derived from the fast, oblique-lateral flow along the unconformity between the basal sedimentary rock and the granite bedrock (Iwatsuki et al., 2001).

Microbiological samples were collected using autoclaved and vacuumed 250-ml stainless steel bottles. The exterior of the sample bottle was wiped with 70% ethanol immediately before entry into the borehole. The groundwater samples were recovered aseptically and used for subsequent geochemical and microbiological measurements (Tables 1 and 2).

Groundwater samples were 0.2 μm-filtered for determination of chemical and isotopic compositions such as Na^+, total Fe (ΣFe), HS^-, SO_4^{2-}, $\delta^{34}S$ of SO_4, NO_3^-, NO_2^-, CO_3^{2-}, HCO_3^-, and dissolved organic carbon (DOC). pH and ^{14}C-age were determined with non-filtered samples. Concentrations of the chemical species were quantified by ion chromatography (DX-120, Dionex, Osaka, Japan), atomic absorption spectrometry (Z-8100, Hitachi, Ibaraki, Japan) and inductively coupled plasma spectrometry (SPS 1200A, Seiko I & E, Chiba, Japan) (Iwatsuki and Yoshida, 1999b).

Groundwater samples were also used for microbiological purposes including the assessment of total cell counts and viable counts as reported in this study (Tables 3 and 4). In addition, the groundwater samples were used for microfloral community analyses based on the 16S rRNA and dissimilatory sulfite reductase genes, which is to be reported separately (Murakami et al., in preparation).

2.3. Total cell counts

Microbiological samples for total counts were collected on 3 different occasions (Table 3) and fixed with borate-buffered formalin at the final concentration of 5% w/v. From 0.1 – 1 ml of groundwater, microbial cells were collected in triplicate by filtration on pre-blackened Isopore filters (Millipore; pore size, 0.2 μm; diameter, 25 mm) and stained separately with the following four fluorescent dyes: 0.05% w/v acriflavine (AF, Sigma) in distilled water for 3-5 min; 0.05% w/v acridine orange (AO, Sigma) in distilled water for 3-5 min; 0.005% (50 μg ml^{-1}) of 4',6'-diamino-2-phenylindol (DAPI, Sigma Chemical Co.) in distilled water for 3-5 min; and 0.05% w/v SYBR Green I (SYBR. Molecular Probes, Inc.) in distilled water for 15 min. The stained cells were air-dried and counted by epifluorescence microscopy (EFD-3, Nikon) with ultraviolet and blue light excitation. The number of cells per microscopic field was adjusted to be 30-50 (Kirchman et al. 1982) by changing the volume of filtration. A minimum of 10 microscopic fields per filter was examined for cell counting.

2.4. Viable counts by fluorescence staining

Viability of the microorganisms in the groundwater samples was determined based on: 1) the cell membrane integrity in June 1999, 2) the electron transport system (ETS) activity in June 1999, and 3) the esterase activity in December 2000, using fluorescent dyes (Table 6).

The cell membrane integrity was examined by epifluorescence microscopy using the *Bac*Light Live/Dead Kit (Molecular Probes, Inc.). The kit contains the mixture of two fluorescent dyes having different permeability towards intact (live) and damaged (dead) cell membranes. The "live" and "dead" were distinguished by green and red fluorescence, respectively, by blue light excitation (Naganuma, 1996; Naganuma and Miura, 1997). Viability using this kit was thus determined as the percentage of the number of green intact cells to the average total counts in 1999. The groundwater was incubated for 15 min in the dark at room temperature.

The ETS-positive cell numbers were also counted by epifluorescence microscopy. ETS-positive cells would convert the tetrazolium salt, 5-cyano-2,3-ditolyl-tetrazolium chloride (CTC), to the red fluorescent CT-formazan and were thus distinguished from the non-fluorescent cells (Rodriguez et al., 1992; Sherr et al., 1999). Microbial cells in the Tono groundwaters were stained with CTC following the previously published protocols

(Rodriguez et al. 1992; Schaule et al. 1993). The ratio of the ETS-positive cell counts to the average total counts in 1999 was regarded as the CTC-based viability. The groundwater was incubated for 4 hours in the dark at room temperature.

The esterase-positive cell numbers were also counted by epifluorescence microscopy. Esterase positive cells would hydrolyze fluorescein diacetate (FDA) to release fluorescein that fluoresces green by blue-light excitation. Microbial cells were stained with FDA according to Tsuji et al. (1995). The ratio of the esterase-positive cell counts to the average total counts in 2000 was regarded as the FDA-based viability. The groundwater was incubated for 15 min in the dark at room temperature.

2.5. Viable cell number estimation by BART

Viable cell numbers based on culture technique were estimated by the Biological Activity Reaction Tests (BART™) in June 1998 (Table 6). BARTs are often used for quick and approximate assessment of various microbial abundances based on the aggressiveness of the target microbial activities (Haveman, 1996). Three different BART kits for total aerobic bacteria (TAB-BART), sulfate-reducing bacteria (SRB-BART), and denitrifying bacteria (DN-BART) were used. Each BART tube contains a ready-mix medium to be added with a 15-ml sample fluid, leaving 10 ml of headspace that would have had 2 ml O_2, or 2.9 mg O_2. TAB-BART tubes were added with 15 ml each of sample fluids. However, in this study, 25 ml each of sample fluid was added to the BART tubes of SRB and DN to fill the 10 ml-headspace and facilitate anaerobic microbial growth. Some additional TAB-BART tubes were filled with 25 ml of groundwater samples for anaerobic heterotrophic growth, but the results were similar to those prepared with 15 ml for aerobic growth (>5 x 10^6 cells ml^{-1}; data not shown).

Ingredients of the BART media, according to the material safety data sheets, are as follows:
TAB-BART. 59.175 mg (per tube) of glycerol, 59.175 mg of peptone, 59.175 mg of tryptone, 29.925 mg of $Na_2S_2O_3 \cdot 5H_2O$, 8.775 mg of $MgSO_4 \cdot 7H_2O$, and 8.775 mg of K_2HPO_4; pH 7.0 when dissolved in 15 ml distilled water.
SRB-BART. Na-lactate, $Na_2S_2O_3 \cdot 5H_2O$, $MgSO_4 \cdot 7H_2O$, NH_4Cl, $CaSO_4 \cdot 2H_2O$, yeast extract, K_2HPO_4, and $FeSO_4 \cdot 7H_2O$; pH 7.0.
DN-BART. NaCl, KNO_3, and tryptone; pH 7.2.
Further chemical composition data of the SRB- and DN-BART ingredients were not available due to non-disclosure policies of the manufacture.

A 15- or 25-ml aliquot of an undiluted sample was added to a BART tube and incubated in the dark at room temperature for 2 weeks. Each BART tube contained a plastic float sphere having the diameter slightly smaller than the tube's inner diameter. The float sphere served to minimize the diffusion of oxygen into the sample, creating a vertical oxygen gradient in the liquid culture. Visible changes in color, turbidity, precipitation, bubbling and biofilm formation were monitored daily to assess the "aggressiveness" of corresponding microbial activities. The "aggressiveness" was ranked to estimate the approximate abundance of the target microbial population in the groundwater samples. The DN-BART estimate was only expressed as the aggressiveness due to the technical limitation. The result of each BART test was run in triplicate, and the results were expressed as the averages of these triplicates.

2.6. Aerobic and anaerobic MPNs.

Viable counts were also determined as the most probable numbers (MPNs) of aerobic/anaerobic heterotrophs, aerobic/anaerobic sulfur-oxidizers, anaerobic sulfate-reducers, anaerobic nitrate reducers, and anaerobic denitrifiers. The MPNs were determined in June 1999 (Table 6).

The MPNs were estimated by incubating one tenth dilutions of groundwater samples up to a dilution of 10^{-7}. Quintiplicates of the sample waters were diluted with the filtered and autoclaved groundwater that had been pre-collected. Each 1 ml of original or diluted samples was added to 9 ml of a liquid medium. Chemical compositions of the liquid media (per 1000 ml) were as follows:

Aerobic and anaerobic heterotrophs (modified from the LB medium with glucose). 5.0 g of peptone, 0.25 g of yeast extract, and 0.1 g of glucose; pH 7.0.

Aerobic and anaerobic (autotrophic) sulfur-oxidizers I (modified from the *Thiobacillus* medium; Atlas, 1997). 10.0 g of $Na_2S_2O_3 \cdot 5H_2O$; 0.5 g of NH_4Cl, 0.25g of K_2HPO_4, 0.25g KH_2PO_4, and 0.25 g $MgSO_4 \cdot 7H_2O$; pH 7.0.

Aerobic and anaerobic (autotrophic) sulfur-oxidizers II (modified from the *Thiobacillus* medium; Atlas, 1997). 10.0 g of $Na_2S_2O_3 \cdot 5H_2O$, 5.0 g KNO_3, 1.8 g of KH_2PO_4, 1.2g of Na_2HPO_4, 0.5 g of $NaHCO_3$, 0.1 g of $MgSO_4 \cdot 7H_2O$, 0.1 g of $(NH_4)_2SO_4$, 0.03 g of $CaCl_2 \cdot 2H_2O$, 0.02 g of $FeCl_3 \cdot 6H_2O$, and 0.02 g of $MnSO_4 \cdot 7H_2O$; pH 7. This medium contained 74 μM of Fe(III) and could stimulate anaerobic oxidation of thiosulfate by Fe(III).

(Anaerobic) sulfate-reducers (modified from the Postgete's medium E; Atlas, 1997). 3.5 g of Na-lactate, 2.0 g of $MgSO_4 \cdot 7H_2O$, 1.0 g of yeast extract, 1.0 g of NH_4Cl, 1.0 g of Na_2SO_4, 0.5 g of KH_2PO_4, 0.2 g of $FeSO_4 \cdot 7H_2O$, 0.1 g of $CaCl_2 \cdot 2H_2O$, 0.02 g Na-ascorbate, and 0.1 ml of 0.1% Na-resazurin solution; pH 7.3.

(Anaerobic) nitrate reducers and denitrifiers (modified from the nitrate broth; Atlas, 1997). 5.0 g of KNO_3, 5.0 g of peptone, 1.7 g of agar, 0.05 g of KH_2PO_4, 0.05 g of $FeSO_4 \cdot 7H_2O$, and 10 ml of 1% bromothymol blue solution; pH 7. Nitrate reduction was checked by blue color (shift to alkaline pH) of the medium, and denitrification was monitored by both a blue color and bubble formation in the medium.

After 30 days of incubation at room temperature (23°C) in the dark, the MPNs were estimated based on the numbers of positive-growth test tubes at each dilution level.

Statistics. Analysis of variance (ANOVA; Sokal and Rohlf, 1981) was conducted to test the degree of variation among the total counts stained with different fluorescence dyes by using the statistical software package SPSS (SPSS, Inc.). In addition to ANOVA, the correlation coefficients and probability levels of significance (Sokal and Rohlf, 1981) were also calculated by using the SPSS.

3. Results

Table 1
Hydrogeological characteristics of the aquifers and rocks at the sampling depths

	MP System (April-November 1999)			Rock core (1989)[a]	
Depth[b] (m)	Water temperature (°C)	Interstitial water pressure (kgf cm^{-2})	Hydraulic water head[c] (m)	Lugeon unit (l min^{-1} m^{-1} 10 kgf^{-1} cm^{-2})	Permeability coefficient (cm s^{-1})
104	17.2 - 17.4	3.40 - 3.82	66.6 - 70.7	3.54×10^{-3}	4.60×10^{-8}
132	18.4 - 18.6	6.23 - 6.77	65.6 - 70.9	1.72×10^{-2}	2.23×10^{-7}
153	19.0 - 19.2	7.85 - 8.64	68.3 - 76.1	1.39×10^{-2}	1.80×10^{-7}
177	19.4 - 20.1	10.38 - 11.23	66.9 - 75.3	1.73×10^{-1}	2.24×10^{-6}

[a]Power Reactor and Nuclear Fuel Development Corporation, unpublished.
[b]Meters below ground level.
[c]Meters below ground level.

3.1. Geochemistry

The groundwater geochemical profiles taken from the cased, packed borehole showed both depth-to-depth and time-to-time variations in parameters (Table 2). The depth-related variation, *e.g.*, the increase of sulfate and the decrease of total organic carbon (TOC) with depth, was likely influenced in part by downward fluid migration from the overlying sedimentary mudstone through the lignite-bearing layer and to the granite bedrock. In contrast, the temporal fluctuation, particularly of sulfate, was likely caused by occasionally increased or decreased lateral flows and fracture flows. All groundwater samples were slightly to moderately alkaline (pH 7.56 to 9.63) and contained a high concentration of sulfate. The groundwater samples from the sedimentary rocks were generally dilute in ionic strength compared with the groundwater from the granite bedrock. Redox potential (Eh) calculated from the sulfate/sulfide ratio and pH was -350 to -332 mV, indicating that the formation waters were anoxic. Temperature recorded in the undisturbed well before packing was within a range from 17°C at 104 m depth to 20°C at 177 m depth.

3.2. Total cell counts

The total cell numbers were counted by epifluorescence microscopy on the samples stained with acriflavine (AF), acridine orange (AO), DAPI and SYBR Green I. As the ultraviolet excitation for DAPI fluorescence also caused mineral autofluorescence, the use of AO is generally recommended for counting total cell numbers in core samples (Fry 1988). However, groundwater samples contained only a few mineral particles that would autofluorescence, and thus we were able to count DNA-bearing particles by DAPI as well as AO. In addition, AF was used to stain microbial cells in the groundwater, as AF was previously used to stain planktonic protists and bacteria simultaneously, *i.e.*, less specifically (e.g. Naganuma et al.,

1998; Kimura et al, 1999).

The total cell counts in the Tono TH-6 groundwater were within the range from 3.9 x 10^5 to 8.4 x 10^6 cells ml^{-1} (Table 3). This is higher than 1.54 x 10^5 cells ml^{-1} in the previous report (Whitman et al., 1998), and corresponds to the highest numbers observed in Fennoscandian igneous rock aquifers (Pedersen, 2001). Pedersen (2001) states that "the only measured parameter which correlated with the total number was the amount of total organic carbon (TOC). It appears that the total number of microorganisms in groundwater does not correlate with any other measured parameter." Regarding total microbial numbers and groundwater parameters in Tono, the total number of microorganimsms in the groundwater did not correlate with any measured parameter. In addition, even the total count did not apparently correlate with the TOC concentration or the ratio of TOC to inorganic carbon (IC) concentration, possibly an indicator of TOC availability (Table 4).

There was no consistent correlation between the total counts and depth. Relatively low numbers were found at the depth of 132 m (average 0.96 x 10^6 cells ml^{-1}). However, no clear tendency of increase or decrease with depth was seen over the 73 m depth range (104 to 177 m depth). Total cell counts in the granite rock (177 m depth) containing less organic material were comparable to those in the sedimentary rocks (104, 132 and 153 m depth).

Year-to-year variation in the total counts was observed (Table 3). Relatively large variation at the uppermost 104 m sampling depth, could be ascribed to the influence of near-surface fluid flow with seasonal and temporal variation due to rain/snowfall and utilization of water resources. In contrast, total cell counts at the lowermost 177 m sampling depth showed only little year-to-year variation. This small variation suggests that the groundwaters in deeper layers are less subject to temporal influence of the near-surface fluid flow.

The total counts stained with the various dyes were generally well correlated with each other. However, significant variation in cells between the different dyes was observed for certain cases (Table 3). For example, total counts from 104 m depth during the year 2000 showed 6.9-fold variation from 1.21 x 10^6 cells ml^{-1} (AF) to 8.39 x 10^6 cells ml^{-1} (AO). Large 3.0-fold variations observed for 153 m depth (year 2000, AF-AO) and 177 m depth (year 2000, AO-DAPI). Variations of 1.2- to 1.4-fold were observed in the remaining depth. Significant correlations between the dye-specific counts were found for AO-SYBR (probability level of significance, p = 0.014), AO-DAPI (p = 0.026), DAPI-SYBR (p = 0.031), and AF-DAPI (p = 0.032) (Table 5). On the other hand, large variation was also found between some cell counts obtained with different dyes. AF-AO (p = 0.539) and AF-SYBR (p = 0.383). Specificity and variability of AF-staining is discussed later in the discussion section.

3.3. Cell morphology

Most of the cells observed during counting were rod shaped or coccoid, having the size of 0.5 x 1 µm or 0.5 µm in diameter. Little change was found in the major cell morphology and size among the different depths. An exception was that cells with a 1-µm stalk were occasionally observed at 153 m. Stalked microorganisms such as *Caulobacter* are often found in oligotrophic habitats (Poindexter, 1981). Occurrence of stalked cells could suggest that TOC at 153 m is biologically less available both quantitatively and qualitatively. The TOC

Table 2
Geochemical compositions of the groundwater from sediment and granite aquifers

Depth[a] (m)	pH	Na^+ (mg l^{-1})	Mg^{2+} (mg l^{-1})	Ca^{2+} (mg l^{-1})	ΣFe (mg l^{-1})	HS^- (mg l^{-1})	SO_4^{2-} (mg l^{-1})	$\delta^{34}S$[b] (‰)	NO_3^- (mg l^{-1})	NO_2^- (mg l^{-1})	IC[c] (mg l^{-1})	TOC[d] (mg l^{-1})	^{14}C age (years)[f]
	8.8	42.7	0.09	4.73	0.06	16.5	35	3.1	<0.02	<0.02	16.0	-[e]	
104	7.56	59.5	0.13	6.51	0.18	-	0.55	-	<0.02	<0.02	25.1	-	4000
	8.71	62.0	0.14	5.49	0.06	-	17.7	-	0.03	<0.02	25.3	0.4	
	8.7	73	0.1	6.2	0.04	37.4	51	11.4	<0.02	<0.02	26.4	-	
132	8.46	84.5	0.11	6.6	0.05	-	6.38	-	<0.02	<0.02	43.6	0.4	6800
	8.56	84.5	0.12	6.77	0.01	-	6.35	-	<0.02	<0.02	40.0	0.8	
	9.0	45	0.02	2.09	<0.02	15.1	41	6.0	<0.02	<0.02	11.8	-	
153	9.24	43.0	<0.01	1.66	0.02	-	12.1	-	<0.02	<0.02	13.7	0.5	11900
	9.63	44.5	0.01	1.69	0.01	-	12.4	-	<0.02	<0.02	11.3	0.4	
	7.8	46	0.41	15.1	<0.02	<0.1	82	0.8	<0.02	0.04	14.8	-	
177	8.26	52.5	0.12	6.74	0.06	-	50.3	-	<0.02	<0.02	15.7	0.4	13100
	8.24	55.0	0.35	10.7	<0.01	-	48.3	-	<0.02	<0.02	19.2	0.1	

Note. The data of each depth-chemical concentration were taken in June 1993 (upper), June 1999 (middle), and November 1999 (lower).
[a]Meters below ground level.
[b]δ^{34}-SO_4.
[c]Inorganic carbon.
[d]Total inorganic carbon.
[e]Not determined.
[f]^{14}C age corresponds to total carbon as referred in Iwatsuki et al. (2001)

Table 3
Total cell counts (x 10^6 cells ml^{-1}) of groundwater microbial communities

Depth[a] (m)	Fluorescence dye[b] AF	AO	DAPI	SYBR	Average (± SD)
104	1.68 (± 0.04)	-[c]	-	-	4.01 (± 1.50)
	4.43 (± 0.17)	4.29 (± 0.13)	3.20 (± 0.66)	-	
	1.21 (± 0.07)	8.39 (± 2.42)	2.71 (± 0.45)	6.12 (± 2.02)	
132	1.48 (± 0.03)	-	-	-	0.96 (± 0.26)
	0.91 (± 0.05)	0.92 (± 0.08)	1.15 (± 0.19)	-	
	0.57 (± 0.12)	0.77 (± 0.11)	0.62 (± 0.24)	0.39 (± 0.07)	
153	2.25 (± 0.03)	-	-	-	1.53 (±1.05)
	1.16 (± 0.16)	1.14 (± 0.18)	0.97 (± 0.29)	-	
	1.70 (± 0.10)	5.20 (± 2.48)	2.06 (± 0.19)	4.49 (± 0.54)	
177	1.93 (± 0.03)	-	-	-	1.90 (± 0.47)
	2.01 (± 0.17)	1.85 (± 0.35)	1.67 (± 0.46)	-	
	1.51 (± 0.16)	3.01 (± 1.24)	0.97 (± 0.19)	2.71 (± 0.29)	

Note. The data of each depth-dye set were taken in June 1998 (upper), June 1999 (middle), and December 2000 (lower).
[a]Meters below ground level.
[b]AF, acriflavine; AO, acridine orange; DAPI. 4',6'-diamino-2-phenylindol; and SYBR, SYBR Green I.
[c]Not determined.

concentration at 153 m was as low as 0.4-0.5 mg C l^{-1} (Table 2), and contains lignite which is likely composed refractory humics. In addition, TOC concentration was generally 0.5-3.6% of IC concentration which may reflect that easily available organics had already been mineralized. However, we have not had any evidence for the relationship between organic matter of lignite layer and microbial habit. We need to pursue this study to get more information.

3.4. Viable cell counts

The Live counts by *Bac*Light were within the range from a minimum of 4.2 x 10^5 cells ml^{-1} at 132 m depth to 1.8 x 10^6 cells ml^{-1} at 177 m depth (Table 6). On the other hand, viability (% viable/total count ratio) showed a concomitant increase with depth from 32.4% (104 m depth) to 59.2% (177 m depth). It is unclear why higher viability was observed in deeper groundwater. Possible explanation might be that the deeper habitats are more stable without

Table 4
Correlation of total counts (AF, AO and DAPI; x 10^6 cells ml^{-1}), total organic carbon concentration (TOC; mg C l^{-1}) and the ratio of TOC to inorganic carbon (IC; mg C l^{-1}), determined in June 1999. Pearson's correlation coefficient (r) and the probability level of significance (p) were calculated by simple linear regression using SPSS

Liner correlation (Y = a X + b)				r	n	p
X	Y	a	b			
AF	AO	0.962	0.004	0.999	4	0.001
AF	DAPI	0.625	0.418	0.991	4	0.009
AO	DAPI	0.645	0.423	0.990	4	0.010
AF	TOC	7.579	-9.325	0.956	4	0.044
AO	TOC	7.962	-9.522	0.966	4	0.034
DAPI	TOC	12.011	-14.118	0.955	4	0.045
AF	TOC/IC	0.301	-0.3642	0.958	4	0.042
AO	TOC/IC	0.316	-0.372	0.968	4	0.032
DAPI	TOC/IC	0.475	-0.550	0.955	4	0.045
TOC	TOC/IC	0.004	0.007	>0.999	4	<0.001

Table 5
Correlation of the dye-specific total counts (x 10^6 cells ml^{-1}). Pearson's correlation coefficient (r) and the probability level of significance (p) were calculated by simple linear regression using SPSS

Dyes		Y = aX + B		r	n	p
X	Y	a	b			
AF	AO	0.570	2.234	0.257	8	0.539
AF	DAPI	0.577	0.696	0.750	8	0.032
AF	SYBR	3.064	-0.040	0.617	4	0.383
AO	DAPI	0.267	0.816	0.769	8	0.026
AO	SYBR	0.764	0.188	0.986	4	0.014
DAPI	SYBR	0.381	0.284	0.969	4	0.031

Table 6
Viable cell counts and estimates of groundwater microbial communities

	Viable cell count[b] [x 10^5 cells ml^{-1}] (% of total)			BART estimate[c] [culturable cells ml^{-1}]			MPN estimate[d] [culturable cells ml^{-1}]								
							Heterotroph		SOB I		SOB II				
Depth[a] (m)	Live / Dead	CTC	FDA	TAB x 10^6	SRB x 10^3	DN activity	aero x 10^6	anaero x 10^3	aero x 10^4	anaero x 10^4	aero x 10^4	anaero x 10^4	SRB	NRB x 10^4	DNB
104	8.87 (32.4%)	4.13 (10.4%)	0.50 (1.0%)	>5.0	4.2	very aggressive	>2.1	9.3	0.46	0.24	0.25	>2.1	ND[e]	0.075	2
132	4.17 (42.4%)	1.45 (14.6%)	0.61 (9.5%)	>5.0	0.2	moderate	0.46	9.3	2.1	0.24	2.4	2.1	ND	2.1	0.3
153	14.54 (50.8%)	0.19 (1.8%)	0.19 (0.4%)	>5.0	<0.1	moderate	0.46	0.009	2.1	0.24	2.1	0.0009	2.3	>2.1	100
177	17.65 (59.2%)	3.98 (21.5%)	0.19 (0.8%)	>5.0	0.24	very aggressive	2.1	1.5	0.24	2.1	4.3	0.075	ND	>2.1	100

Note. Data of BART estimates were collected in 1998. Data of Live/Dead and CTC counts and MPN estimates were collected in 1999. Data of FDA counts were taken in 2000.

[a]Meters below ground level.

[b]Live/Dead, *Bac*Light Kit (Molecular Probes, Inc.); CTC, 5-cyano-2,3-ditolyl-tetrazolium chloride (substrate for electron transport system-related enzymes); FDA, fluorescein diacetate (substrate for esterases).

[c]TAB, total aerobic heterotrophic bacteria; SRB, sulfate-reducing bacteria; DN, denitrification.

[d]SOB, sulfur-oxidizing bacteria; aero, aerobic; anaero, anaerobic; SRB, sulfate-reducing bacteria; NRB, nitrate-reducing bacteria; DNB, denitrifying bacteria.

[e]not detected.

major environmental changes. It also seems reasonable that the microflorae in deeper groundwaters would be the survivors of long-term migration through rock matrices and thus selected for retaining higher viability.

The CTC-positive cell counts, *i.e.*, ETS-active cell counts, varied from a minimum at the sediment-granite unconformity at 153 m depth of 1.9 x 10^4 cells ml^{-1} (153 m depth) to 4.0 - 4.1 x 10^5 cells ml^{-1} (104 m and 177 m depth) (Table 6).

The minimum viability was also found at 153 m depth. This low frequency of ETS-active cells at 153 m depth also corresponds with the low BART-estimates of sulfate reducers, *i.e.*, sulfate-respiring cells (Table 6). In addition, a respiring strain of the common subsurface genus *Comamonas* is known to be non-stainable with CTC (Fuller et al., 2000), which confirms the previous finding that CTC is not a general and universal dye for staining ETS-active cells (Ullrich et al., 1996). The microbial community at 153 m depth was possibly dominated by fermentative and/or respiring-but-non-stainable species. However, the possibility of fermenters' dominance could be contradicted by the low MPN of anaerobic heterotrophs to which fermentative microorganisms are categorized (Table 6). Therefore, the low CTC-counts at 153 m depth might reflect a low abundance of anaerobic heterotrophs such as sulfate-respiring (sulfate-reducing) bacteria, rather than fermentative bacteria.

The FDA-positive (esterase-active) cell counts varied from 1.5 x 10^4 cells ml^{-1} (153 m depth) to 6.1 x 10^4 cells ml^{-1} (132 m depth) (Table 6). The maximum viability for FDA-positive bacteria, of 9.5% was found accordingly at 132 m depth, while the FDA-based viability was 1% or less and corersponds to a at the other depths. This is one of the few cases that a maximum in cell viability counts was found at 132 m depth; the other example was the maximum in counts for anaerobic heterotrophs. The correlation between the FDA-positive counts and other viable counts was insignificant (Table 7). It is reported that the culturable cell number and the viable counts using FDA and FDA-relatives may not always correlate (Tsuji et al. 1995; Fuller et al., 2000). This discrepancy has not thoroughly been explained, however, the use of FDA and its relatives still provide some information about the metabolic state of subsurface microbial communities.

3.5. Viable count estimates by BART

Each BART test was made in triplicate and showed only little variation among the three tubes. The estimated number of total aerobic bacteria (TAB) was >5 x 10^6 cells ml^{-1} in all the samples. Due to BART's technical limitation, TAB-counts of >5 x 10^6 cells ml^{-1} are regarded as "above the estimation level". The high TAB-estimates exceeded the total counts in Table 3. This could be due to the occurrence of uncounted but viable dwarf cells (Kieft, 2000) in the groundwater. For example, verrucae microbial cells as small as 0.33 x 0.49 μm are known to occur at a density of 10^5 cells g^{-1} dry rice paddy soil (Jansen et al., 1997).

The number of sulfate-reducing bacteria (SRB) estimated by BART was of the order of 10^2 to 10^3 cells ml^{-1} at the depths of 104, 132 and 177 m (Table 6). It is notable that the SRB-estimate by BART at 153 m was below the detection limit. The uniqueness of the 153 m-depth layer regarding the different depth profile of SRB growth with each media is hydrologically supported by the relatively high hydraulic water head (Table 2). From the

Table 7
Correlation of the dye-specific viable counts (x 10^5 cells ml^{-1}). Pearson's correlation coefficient (r) and the probability level of significance (p) were calculated by simple linear regression using SPSS.

Dyes		Y = aX + b		r	n	p
X	Y	a	b			
Live/Dead	CTC	0.051	1.864	0.157	4	0.843
Live/Dead	FDA	-0.030	0.771	-0.951	4	0.049
CTC	FDA	0.018	0.319	0.151	4	0.849

hydraulic water head data, the groundwater in the Lower Toki lignite-bearing Formation below the low permeability layer, at 153 m, may be infiltrated horizontally from the Seto Group along the unconformity between the sedimentary rocks and granite. The ^{14}C-age data also shows the by the downward fluid flow and low permeability between 132 m and 153 m depth (Table 3). This hydrological uniqueness of the 153 m-depth layer should be reflected in the microfloral composition, with particular reference to anaerobic respirers such as SRB, because of the closed flow system.

Denitrification is also a process of anaerobic respiration, and the distribution of denitrification (DN) activity was also estimated by BART. The DN activity was highest for the 104 and 177 m samples, and 'moderate' for the 132 and 153 m-depth samples (Table 6). It is unclear how many DN counts actually account for the 'very aggressive' and 'moderate' DN activity. It should be safe to conclude only qualitatively that DN activity (and possibly DN abundance) was higher in the shallow sediment rock and granite bedrock than in the intermediate mudstone and lignite-bearing layers. Generally, DN activity occurs in a relatively less reducing, suboxic condition close to the redox front (Blackburn and Blackburn, 1992; Nealson, 1997; Lovely, 2001). These conditions are likely found in the shallow 104 m-depth layer that may occasionally have communication with near surface flow. In addition, the 177 m-depth layer was probably less reducing, due to occasional influx of younger, less reducing, fluids along major fractures.

3.6. Viable count estimates by MPN

Culturable subsurface microorganisms in borehole TH-6 were counted by the MPN method based on serial dilution of selective culture media (Table 6). First, aerobic heterotrophs occurred at MPNs ranging from 4.6 x 10^4 (132 and 153 m) to the number over 2.1 x 10^6 cells ml^{-1} (104 and 177 m depth). The high aerobic heterotroph MPN counts (x10^6 cells ml^{-1}) often exceed the total cell counts in Table 3. This could be due to the occurrence of uncounted but viable dwarf cells (Kieft, 2000), as similarly seen for TAB-BART estimates.

The MPN of anaerobic heterotrophs was as low as 9.0 x 10^0 (153 m depth) to 9.3 x 10^3 cells ml^{-1} (104 and 132 m). This depth-profile was similar to that for anaerobic sulfur-oxidizers (II), and showed no apparent correlation with depth or with other chemical parameters, including TOC concentration. It is likely that this heterogenous distribution of the Tono subsurface microorganisms may be influenced locally by fracture-controlled hydrology and non-steady-state geochemical conditions.

The MPN counts of autotrophic sulfur-oxidizing bacteria (SOB) were obtained with two types of media (I and II), and showed no consistent result between aerobes and anaerobes with depth (Table 6). There was also no consistency of results obtained with the two types of media. The sulfur-oxidizing MPN counts varied within a range between 9.0 x 10^0 cells ml^{-1} (SOB medium II, anaerobic; 153 m counts) and 4.3 x 10^4 cells ml^{-1} (SOB medium II, aerobic; 177 m depth). This range accounted for <0.001% and 2.3%, respectively, of the average (of AF-, AO- and DAPI-based) total counts of each corresponding depth determined in 1999. The SOB medium II contained 74 µM of Fe(III) which could oxidize thiosulfate, and presumably yielded slightly higher SOB-MPN than the SOB medium I. However, the lowest SOB-MPN was also obtained from the medium II culture. In general, there was no consistency among the samples and culture conditions for the SOB-MPN counts, and it is unclear why.

MPN counts for sulfate-reducing bacteria (SRB) were positively determined only at 153 m, where SRB-BART was below the detection limit. There was no observable correlation between SRB-MPN and SRB-BART counts, nor between either of these counts and the concentrations of sulfate or sulfide. It is previously reported that "no correlation could be found between sulfate and sulfide concentrations and SRB" in igneous rock aquifers (Pedersen, 2001), which is also true for this study at borehole TH-6, in the sand-/mudstones and granite. Inconsistency between the two types of media could be due to differences in the chemical composition of each media. Detailed chemical data of the BART-SRB medium is not available.

The MPN counts of nitrate-reducing bacteria (NRB) and denitrifying bacteria (DNB) generally showed similar increases with depth. DN-BART indicated high DN activity ("very aggressive") at the deepest sampling depth (177 m depth); however, it also showed a "very aggressive" score at the shallowest sampling depth (104 m depth). The depth-profiles of DN and DNB were not well correlated to, nor explained by, other microbiological and geochemical parameters. The NRB-MPN corresponded to >1 to >2% of the average (of AF-, AO- and DAPI-based) total counts of each corresponding depth determined in 1999. In contrast, the DN-MPN counts at most accounted for only 0.065% of the total counts (153 m).

4. Discussion

4.1. Distribution of total cell counts

Granite bedrock is generally poor in organic materials. However, it should be noted that the total cell counts in the organic-poor Tono granite bedrock (177 m depth), as shown by low TOC in Table 2, were comparable to those in the sedimentary rocks (104, 132 and 153 m depth) (Table 3). Abundance of groundwater microbial communities did not much change from the 104 m-depth sedimentary rock to the 177 m-depth granite bedrock. That may be explained by migration of water-borne microorganisms via downward fluid flow, entraining lignite-derived organics from intermediate layers. Although ^{14}C-age indicates the occurrence of downward fluid flow, the groundwater potential, i.e., hydraulic water head, suggests the possibility of upward flow from 177 m depth to 153 m depth (Table 1). However, the impermeable layer (permeability of $<10^{-8}$ m s^{-1}; Iwatsuki et al., 2001) lying in-between would virtually block the potential upward flow. In other words, blockage by the impermeable layer causes the higher groundwater pressure and higher water head, accordingly. Locally focused fracture-flow could migrate at relatively high velocity, as shown by the 1600-year increase in the ^{14}C-age for a 24-m depth increase over the sediment-granite unconformity (153 m to 177 m depth). It should be noted that the error involved in the ^{14}C-age determination is ±1500 years and that the 1600-year difference in ^{14}C-age may be much smaller. In fact, other studies suggest that there would have been inflow of relatively 'young' fluid along the fractures crossing the borehole TH-6 near the depth of 177 m, which may have 'juvenilized' the apparent ^{14}C-age of the 177 m-depth fluid (Iwatsuki et al., 1999, 2001). The apparent downward migration rate, 15 mm y^{-1}, is higher than those of 10 and 4 mm y^{-1}, calculated based on the ^{14}C-age and depth ranges of 28 m/2800 years (104-132 m depth) and 21 m/5200 years (132-153 m), respectively (Table 2). These rates are low enough to consider that the depths investigated in this study are within the 'intermediate flow system', which are less connected with the surface 'local flow system' (Lovely and Chapelle, 1995).

Downward flow, porous or fracture-controlled, may transport microorganisms and organic matter to the upper granite layer without significant loss of microbial viability and organic matter, but possibly with some modification of microfloral composition (Lehman et al., 2001). Although microbial migration is rather limited due to adhesion to mineral surfaces (Fletcher and Murphy, 2001), the migrant bacteria should be able to populate the granite layer after enough doubling times. The microbial communities in the upper granite layer would be supported by allochthonous input from the overlying sedimentary rocks, as well as by possible autochthonous production via microbial chemolithoautotrophy and chemolithoheterotrophy (Amy and Haldeman, 1997).

4.2. Dye-to-dye variation in total cell counts

Dye-specific total counts showed significant variation in certain cases, although dye-specific total counts were generally well correlated among the dyes (Table 3). Specifically, AF-AO-DAPI in 1999 showed a highly significant correlation (p 0.01, Table 4). Generally, significant correlations ($p < 0.05$) were found for AF-DAPI, AO-DAPI, AO-SYBR and DAPI-SYBR (Table 5). An ideal correlation would show a relationship close to 1:1 between the two dye-specific counts, resulting in the slope and y-intercept of 1 and 0 (**a** and **b** in Table 5), respectively. The AO-SYBR relationship in general was the closest to the ideal correlation having a slope near-1 and y-intercept near-0, with the highest correlation efficient ($r = 0.986$). Other correlations were remote from the ideal relationship, resulting in slopes and y-intercepts far from 1 and 0, respectively, despite high r-values.

The AO-counts were generally higher than DAPI- and SYBR-counts (Table 3), in agreement with the previous report of high-AO and low-DAPI tendency (Harvey, 1987). However, the AO-counts tended to be lower than AF-counts (Table 3), which disagrees with the previous report of high-AO and low-AF tendency (Bergström et al., 1986). SYBR Green I also yields higher counts than DAPI (Weinbauer et al., 1998) but resulted in slightly lower counts than AO (Table 3).

If drawing linear correlation graphs, it may be noticed that one or two 'outlying' data lowered the whole correlation and that the correlations without those 'outliers' would have been much closer to ideal. Examples of the outliers would be high-AO (and high-SYBR) and low-AF (and low-DAPI) observed in 2000. The occasional discrepancy in total counts between the dyes is not fully elucidated, and further systematic examination is needed.

4.3. Dye-to-dye variation in viable counts

The viable counts measured by Live/Dead dyes, which are based on cell membrane integrity, showed an overall increase with depth, while CTC- and FDA-counts based on enzymatic activities showed different tendencies (Tables 6 and 7). The reason why Live/Dead-counts increased with depth is not clear. One might expect that the viability, whatever the definition is based on, decreases with depth as physical constraints such as reverses of porosity and permeability increase (Colwell, 2001). However, higher Live/Dead-counts in deeper zones may suggest that the habitat conditions such as temperature and fluid flow are more stable and thus allows more structural stability of microbial cells.

The CTC-counts based on enzymatic activity involved with the electron transport system, (*i.e.*, respiration), decreased with an increase in depth to 153 m depth. However, high CTC-counts and the highest CTC-viability (ratio of CTC-counts to total) were observed at 177 m depth (Table 6). The minimum CTC-viability of 1.8% at 153 m depth could be explained by the dominance of respiring but non-stainable cells such as a *Comamonas* strain (Fuller et al., 2000). The idea of 'respiring but non-stainable' (RBNS) is closely similar to the idea of 'viable but non-culturable' (VBNC) and should be re-considered in relevance to biogeochemistry and microbial syntrophism (Grimes et al., 2000). The microbial community at 153 m depth was possibly dominated by anaerobically respiring (*e.g.*, sulfate-respiring) RBNS bacteria and/or by fermentative VNBC bacteria, as shown by the low but positive SRB-MPN counts and the very low MPN-counts of anaerobic heterotrophs, the latter including fermentative bacteria (Table 6).

The FDA-counts tended to decrease with the increase in depth, which was opposite to the depth-profile of Live/Dead-counts (negative correlation, Table 7). However, the decrease in FDA-viability was not perfectly consistent with depth. The maximum FDA-viability, 9.5%, was seen at 132 m depth; the minimum, 0.4%, was found at the underlying depth of 153 m. The high FDA-viability at 132 m depth could imply that microorganisms actively mediate the enzymatic erosion of lignite-related organic materials in that layer. Relatively high IC concentration at 132 m depth (Table 2) suggests that lignite-related organics are rapidly eroded and mineralized, leaving only little un-mineralized TOC. By contrast, the low FDA-viability at 153 m depth was presumably attributed to limited bioavailability of buried organic matter in the bottom layer of this sedimentary regime. Labile organic matter must have been mineralized already over geological time, with only refractory organics now remaining. This might explain other low scores for CTC-viability, BART-SRB, anaerobic heterotrophic MPN and anaerobic SOB-MPN (Table 6), although general correlations among the scores were inconsistent. FDA-counts and culturable cell numbers such as MPN counts may not always correlate with each other (Tsuji et al. 1995; Fuller et al., 2000). The methodology-based discrepancy in viable counts is not fully elucidated, and more systematic examinations are needed.

4.4. Distribution of viable count estimates by BART and MPN cultures

A number of viable cell counting methods using selective culture media are available for the microorganisms involved in target biogeochemical processes. This study focused on including both fermentative and respiring heterotrophs, autotrophic sulfur-oxidizing bacteria (SOB), heterotrophic sulfate-respiring (sulfate-reducing) bacteria (SRB), heterotrophic nitrate-respiring (nitrate-reducing) bacteria (NRB), and denitrifying bacteria (DNB). It has been said that a correlation of some viable counts with other counts would fail (e.g., Pedersen, 2001), and we faced similar difficulties in finding consistent trends between the viable count estimates by BART and MPN cultures (Table 6). The incoherence among the data may imply that each sampled layer has a unique set of microflora utilizing a different set of electron donors and acceptors in the groundwater.

Geological and geochemical characterization of the study area is needed to design a more systematic sampling program in terms of depths and time intervals. Selection of methods and target organisms is also more seriously considered. For example, future studies should include more extensive measurements of Eh, HS^-, TOC, etc., and more complete total and viable counting with various dyes and cultures. In addition, we should have grown TAB-BART cultures anaerobically. More time and attention will be given to all of these measurement in future and ongoing programs of the Tono subsurface microbiological study.

4.5. Microbial activity relevant to S-cycling

Oxidation and reduction of sulfur is one of the dominant biogeochemical processes in subsurface environments (Grossman and Desrocher, 2001; Nealson, 1997). Our results (Table 6) show the occurrence of culturable S-oxidizing and S-reducing bacteria (SOB and SRB, respectively) throughout the studied depths, with high and low numbers according to culture conditions. Although their activities were not measured in this study, their *in situ* activities may be recorded in geological media. For example, active S-reduction at 132 m is suggested by a $\delta^{34}SO_4^{2-}$ value as high as 11.4‰, compared with lighter values of +0.8 to +6.0‰ at other depths (Table 2). On the other hand, lighter $\delta^{34}SO_4^{2-}$ values may be the result of S-oxidation of H_2S with a lower $\delta^{34}SO_4^{2-}$ value. Thus, the light $\delta^{34}SO_4^{2-}$ values suggest possible S-oxidation at 104 and 177 m depth, which was also observed in the general distribution of SOB (Table 6).

High sulfate-reduction at 132 m, as inferred by the high $\delta^{34}SO_4^{2-}$ value may correspond to the high CTC-viability at this depth (14.6%, Table 6). This sulfate reduction could be coupled to oxidation of organic matter in the lignite-bearing layer. However, the SRB-BART and SRB-MPN were not particularly high at 132 m depth, as the SO_4^{2-} concentration was the lowest at this depth (Table 6). This is an example of no visible correlations between the concentrations or isotopic signatures of sulfate and sulfide concentrations and SRB (Pedersen, 2001) and of isotopic signatures as indicators of microbial activity (Des Marais, 1996).

The $\delta^{34}S$ value of pyrite often provides a signature for microbial sulfate reduction in the geological record. Framboidal pyrite has previously been observed in the Akeyo formation, corresponding to our sampling depths of 104 and 132 m. Euhedral to subhedral pyrite was abundantly found in the lower Toki formation, corresponding to our 153 m sampling depth. The $\delta^{34}S$ value of the framboidal pyrite was as low as -14 to -8‰, whereas the euhedarl and subhedral pyrite showed a higher $\delta^{34}S$ values of +10 to +43‰ (Shikazono and Utada, 1997). It is generally assumed that a low (or light) $\delta^{34}S$-pyrite value reflects biogenic sulfate reduction. In addition, framboidal and eu-/subhedral pyrites are likely biogenic and abiogenic, respectively. Therefore, the low- $\delta^{34}S$ of framboidal pyrite at 104 and 132 m depth and the high- $\delta^{34}S$ of euhedral or subhedral pyrite at 153 m depth are likely to have biogenic and abiogenic origins, respectively. This inference of microbial sulfate reduction at 104 and 132 m depth coincides with the profiles of CTC-viability and SRB-BART scores, but contradicts the SRB-MPN profile that showed positive S-reduction only at 153 m (Table 6). This does not necessarily refute the reliability of SRB-MPN estimation, and could be related to the occurrence of archaeal sulfate reducers specifically at 153 m depth (Murakami et al., 2000, in preparation).

Sulfate reducing bacteria (SRB) may mediate uranium reduction either coupled or uncoupled with growth (Ehrlich, 1996; Lloyde and Macaskie, 2001). The geological setting for uranium deposit formation in the Tono area is explained as follows: granitic U(VI) dissolved in groundwater is transported to the basal layer of the Toki lignite sediment regime where sulfate reduction likely occurs. Coupled with microbial sulfate reduction, dissolved granitic U(VI) is reduced to less soluble U(IV) and precipitated to form uranium deposits in the sediment. The coupling of microbial sulfate reduction and uranium deposition is suggested by the co-occurrence of uranium and framboidal pyrite (Japan Nuclear Cycle Development Institute, unpublished). In addition, calcite serves as a sink for uranium in depth groundwater aquifers (Struchio et al., 1998), and is one of the major fracture-filling minerals in the Tono area (Iwatsuki and Yoshida, 1999b). SRB mediate the precipitation of High-Mg calcite and Ca-dolomite (Vasconcelos et al., 1995), and the presence of biomolecules such as aspartic acid enhances the growth of calcite (Teng et al., 1998). Thus, it is likely that SRB may facilitate the uranium deposit formation by direct U(VI) reduction and indirect U(IV) immobilization via calcite precipitation. The subsurface SRB activity in the Tono uranium deposit area should be evaluated regarding its importance for the immobilization of U over geologic time frames.

5. CONCLUSIONS

In this study, it was suggested that microbial abundance has been influenced by hydrological constraints rather than geochemical conditions. By contrast, microbial diversity and activity showed depth profiles. In particular, microbial activities were relatively high at 132 m depth, and were remarkably low at 153 m depth. However, geochemical influences for the depth-related microbial diversity and activity were obscure. Long-term monitoring of microbial parameters is needed to evaluate the data collected at scattered time points in the context of long-term fluctuation.

Acknowledgments

We are grateful to the invaluable help from Tono Geoscience Center, Japan Nuclear Cycle Development Institute (JNC). We also thank Dr. Kevin Mandernack for reading the manuscript. This work was supported by: the Cooperative Research Scheme on the Nuclear Fuel Cycle of JNC; the REIMEI Research Resource of Japan Atomic Energy Research Institute; and a grant from Japan National Oil Corporation; and the Special Coordination Fund "Archaean Park Project" from the Ministry of Education, Culture, Sports, Science and Technology (MEXT) of Japan.

REFERENCES

Amy, P. S., and D. L. Haldeman (eds.) *Microbiology of the Terrestrial Deep Subsurface.* CRC Press, Boca Raton, Florida, 1997.

Atlas, R.M. *Handbook of Microbiological Media (2nd ed.)* CRC Press, Boca Raton, Florida, 1997.

Bergström, I., A. Heinanen, and K. Salonen. Comparison of acridine orange, acriflavine, and bisbenzimide stains for enumeration of bacteria in clear and humic waters, *Appl. Environ. Microbiol.*, 51, 664-667, 1986.

Blackburn, T. H., and N. D. Blackburn. Model of nitrification and denitrification in marine sediments, *FEMS Microbiol. Ecol.*,100, 517-522, 1992.

Colwell, F. S. Constraints on the distribution of microorganisms in subsurface environements, in *Subsurface Microbiology and Biogeochemistry*, edited by J. K. Fredrickson and M. Fletcher, pp. 71-95, Wiley-Liss, Inc., New York, 2001.

Des Marais, D. J., Stable light isotope biogeochemistry of hydrothermal systems, in *Evolution of Hydrothermal Ecosystems on Earth (and Mars?)*, edited by G. R. Bock and J. A. Goode, pp. 83-98, John Wiley & Sons, Ltd., Chichester, 1996.

Ehrlich, H. L., *Geomicrobiology. (3rd ed.)*, pp. 719, Marcel Dekker, Inc., New York, 1996.

Fletcher, M., and E. Murphy, Transport of microorganisms in the subsurface: the role of attachment and colonization of particle surfaces, in *Subsurface Microbiology and Biogeochemistry*, edited by J. K. Fredrickson and M. Fletcher, pp. 39-68, Wiley-Liss, Inc., New York, 2001.

Floyd, J. S., J. C. Mutter, A. M. Doodliffe, and B. Taylor, Evidence for fault weakness and fluid flow within an active low-angle normal fault, *Nature*, 411, 779-783, 2001.

Fredrickson, J. K., and T. J. Phelps, Subsurface drilling and sampling, in *Manual of Environmental Microbiology*, edited by C. J. Hurst, G. R. Knudsen, M. J. McInerney, L. D. Stetzenbach and M. V. Walter, pp. 526-540, ASM Press, Washington, DC, 1997.

Fredrickson, J. K., J. P. McKinley, B. N. Bjornstad, D. B. Ringelberg, D. C. White, L. R. Krumhols, J. M. Suflita, F. S. Colwell, R. M. Lehman, and T. J. Phelps, Pore-size constraints on the activity and survival of subsurface bacteria in a late Cretaceous shale-sandstone sequence, northwestern New Mexico, *Geomicrobiol. J.*, 14, 183-202, 1997.

Fredrickson, J. K., J. M. Zachara, D. W. Kennedy, H. Dong, T. C. Onstott, N. W. Hinman, and S. Li, Biogenic iron mineralization accompanying the dissimilatory reduction of hydrous ferric oxide by a groundwater bacterium, *Geochim. Cosmochim. Acta*, 62, 3239-3257, 1998.

Fry, J. C., Determination of biomass, in *Methods in Aquatic Bacteriology*, edited by B. Austin, pp. 27-72, Wiley, Chichester, 1988.

Fuller, M. E., S. H. Streger, R. K. Rothmel, B. J. Mailloux, J. A. Hall, T. C. Onstott, J. K. Fredrickson, D. L. Balkwill, and M. F. DeFlaun, Develpment of a vital fluorescent staining method for monitoring bacterial transport in subsurface environments, *Appl. Environ. Microbiol.*, 66, 4486-4496, 2000.

Garland, G. D., *Introduction to Geophysics: Mantle, Core, and Crust (2nd ed.)*. Saunders, Philadelphia, 1979.

Gold, T., The deep, hot biosphere, *Proc. Natl. Acad. Sci. USA.*, 89, 6045-6049, 1992.

Grimes, D. J., A. L. Milles, and K. H. Nealson, The importance of viable but nonculturable bacteria in biogeochemistry, in *Nonculturable Microorganisms in the Environment.*, Edited by R. R. Colwell and D. J. Grimes, pp. 209-227, ASM Press, Washington, DC, 2000.

Grossman, E. L., and S. Desrocher, Microbial sulfur cycling in terrestrial subsurface environments, in *Subsurface Microbiology and Biogeochemistry*. edited by J. K. Fredrickson and M. Fletcher, pp. 219-248, Wiley-Liss, Inc., New York, 2001.

Harvey, R. W., A fluorochrome-staining technique for counting bacteria in saline, organically enriched, alkaline lakes, *Limnol. Oceanogr.*, 32, 993-995, 1987.

Haveman, S. A., S. Stroes-Gascoyne, C. J. Hamon, and C. L. Messerer, Microbial analysis of water collected from privately owened wells and a beaver pond. Atomic Energy of Canada Ltd., Technical Report TR-757, COG-96-357-1, 1996.

Hazen TC, L. Jimenez, G. L. de Victoria, and C. B. Fliermans, Comparison of baccteria from deep subsurface sediment and adjacent groundwater, Microb. Ecol., 22, 293-304, 1991.

Iwatsuki, T., and H. Yoshida, Characterizing the chemical containment properties of the deep geosphere: water-rock interactions in relation to fracture systems within deep crystalline rock in the Tono area, Japan, in *Chemical Containment of Waste in the Geosphere*. eds. R. Metcalfe and C. A. Rochelle, pp. 71-84. London: Geological Society, 1999a.

Iwatsuki, T., and H. Yoshida, Groundwater chemistry and fracture mineralogy in the basement granitic rock in the Tono uranium mine area, Gifu Prefecture, Japan:

Groundwater comsposition, Eh evolution analysis by fracture filling materials. *Geochem. J.*, 33, 19-32, 1999b.

Iwatsuki, T., K. Hama, and H. Yoshida, Geochemical investigation of groundwater in the Tono area, Japan. PNC Technical Review 102: 31-37 (in Japanese with English abstract). Power Reactor and Nuclear Fuel Development Corporation (PNC; reformed in 1998 as Japan Nuclear Cycle Development Institute, JNC), Tokai, Ibaraki, Japan, 1997.

Iwatsuki, T., S. Xu, S. Ito, M. Abe, and M. Watanabe, Corroboration of hydraulic conditions of groundwater using C-14. JNC Technical Review 4: 93-100 (in Japanese with English abstract). Japan Nuclear Cycle Development Institute, Tokai, Ibaraki, Japan, 1999.

Iwatsuki, T., S. Xu, Y. Mizutani, K. Hama, H. Saegusa, and K. Nakano, Carbon-14 study of groundwater in the sedimentary rocks at the Tono study site, central Japan. Applied Geochemistry, 16, 849-859, 2001.

Jansen, P. H., A. Schuhmann, E. Morschell, and F. A. Rainey, Novel anaerobic ultramicrobacteria belonging to the *Verrucomicrobiales* lineage of bacterial descent isolated by dilution culture from anoxic rice paddy soil. *Appl. Environ. Microbiol.,* 63, 1382-1388, 1997.

Japan Nuclear Cycle Development Institute, Project to Establish Scientific and Technical Basis for HLW Disposal in Japan: Project Overview Report. Supplement Report 1. JNC TN1410 2000-002, 2000.

Kieft, T. L, Size matters: dwarf cells in soil and subsurface terrestrial environments, in *Nonculturable Microorganisms in the Environment.* ed. R. R. Colwell and D. J. Grimes, pp. 19-46, Washington, DC: ASM Press, 2000.

Kimura, H., T. Fukuba, and T. Naganuma, Biomass of thraustochytrid protoctists in coastal water. *Mar. Ecol. Prog. Ser.,* 189, 27-33, 1999.

Kirchman, D., J. Sigda, R. Kapuscinski, and R. Mitchell, Statistical analysis of the direct count method for enumerating bacteria, *App. Environ. Microbiol.*, 44, 376-382, 1982.

Lehman, R. M., F. S. Colwell, and G. A. Bala, Attached and unattached microbial communities in a simulated basalt aquifer under fracture- and porous-flow conditions, *Appl. Environ. Microbiol.*, 67, 2799-2809, 2001.

Liang, L., N. Korte, B. Gu, R. Puls, and C. Reeter, Geochemical and microbial interactions affecting the long-term performance of in situ 'iron' barriers, *Adv. Environ. Res.*, 4, 273-286, 2000.

Lloyde, J. R., and L. E. Macaskie, Bioremediation of radionuclide-containing wastewaters, in *Environmental Microbe-Metal Interactions,* edited by D. R. Lovely, pp.277-327, ASM Press, Washington, DC, 2000.

Lovely, D. R. (ed.), *Environmental Microbe-Metal Interactions*, ASM Press, Washington, DC, 2000.

Lovely, D. R., Reduction of iron and humics in subsurface environments, in *Subsurface Microbiology and Biogeochemistry*, edited by J. K. Fredrickson and M. Fletcher, pp. 193-217, Wiley-Liss, Inc., New York, 2001.

Lovely, D. R., and F. H. Chapelle, Deep subsurface microbial processes, *Rev. Geophys.*, 33,

365-381, 1995.

Lower, S. K., M. F. Hochella Jr., and T. J. Beveridge, Bacterial recognition of mineral surfaces: nanoscale interactions between *Shewanella* and α-FeOOH, *Science*, 292, 1360-1363, 2001.

Murakami, Y., T. Iwatsuki, and T. Naganuma, Abundance and diversity of the deep groundwater microflora in the Tono Uranium Mine area, central Japan. Eos Trans. AGU, 81 (48), Fall Meet. Suppl., Abstract B71A-02, 2000.

Naganuma, T., Differential enumeration of intact and damaged marine planktonic bacteria based on cell membrane integrity, *J. Aquat. Ecosyst. Health*, 5, 217-222, 1996.

Naganuma, T., and S. Miura, Abundance, production and viability of bacterioplankton in the Seto Inland Sea, *J. Oceanogr.*, 53, 435-442, 1997.

Naganuma, T., H. Takasugi, and H. Kimura, Abundance of thraustochytrids in coastal plankton, *Mar. Ecol. Prog. Ser.*, 162, 105-110, 1998.

Naganuma, T., Y. Murakami, and T. Iwatsuki, Formation of colloidal iron hydroxide by deep subsurface bacteria, International Chemical Congress of Pacific Basin Societies (Pacifichem 2000), 14-19 December 2000, Honolulu, Hawaii, p.40, 2000.

Nealson, K. H., Sediment bacteria: who's there, what are they doing, and what's new? *Annu. Rev. Earth Planet. Sci.*, 25, 403-434, 1997.

Nealson, K. H., and D. Saffarini, Iron and manganese in anaerobic respiration: environmental significance, physiology, and regulation, *Annu. Rev. Microbiol.*, 48, 311-343, 1994.

Pedersen, K., The deep subterranean biosphere, *Earth-Sci. Rev.*, 34, 243-260, 1993.

Pedersen, K., Diversity and activity of microorganisms in deep igneous rock aquifers of the Fennoscandian Shield, in *Subsurface Microbiology and Biogeochemistry*, edited by J. K. Fredrickson and M. Fletcher, pp. 97-139, Wiley-Liss, Inc., New York, 2001.

Pedersen, K., and S. Ekendahl, Distribution and activity of bacteria in deep granitic groundwaters of southeastern Sweden, *Microb. Ecol.*, 20, 37-52, 1990.

Pedersen, K., J. Arlinger, L. Hallbeck, and C. Pettersson, Diversity and distribution of subterranean bacteria in groundwater at Oklo in Gabon, Africa, as determined by 16S-rRNA gene sequencing technique, *Mol. Ecol.*, 5, 427-436, 1996.

Poindexter, J. S., The caulobacters: ubiquitous unusual bacteria, *Microbiol. Rev.*, 45, 123-179, 1981.

Power Reactor and Nuclear Fuel Development Corporation, *Nihon-no-uran-shigen* (Uranium Resources in Japan), PNC TN7420 94-006, pp. 11, 1994.

Press, F., and R. Siever, Earth (4th edition), p 656, W. H. Freeman and Company, New York, 1986.

Rodriguez, G. G., D. Phipps, K. Ishiguro, and H. F. Ridgway, Use of a fluorescent redox probe for direct visualization of actively respiring bacteria, *Appl. Environ. Microbiol.*, 58, 1801-1808, 1992.

Schaule, G., H.-C. Flemming, and H. F. Ridgeway, Use of 5-cyano-2,3-ditolyl tetrazolium chloraide for quantifying planktonic and sessile respiring bacteria in drinking water, *Appl. Environ. Microbiol.*, 59, 3850-3857, 1993.

Sherr, B. F., P. del Giorgio, and E. B. Sherr, Estimating abundance and single-cell characteristics of respiring bacteria via the redox dye CTC, *Aquat. Microb. Ecol.*, 18, 117-131, 1999.

Shikazono N, and M. Utada, Stable isotope geochemistry and diagenetic mineralization associated with the Tono sandstone-type uranium deposit in Japan, *Mineralium Deposita* 32, 596-606, 1997.

Sokal, R. R., and F. J. Rohlf, *Biometry (2nd ed.)*. W. H. Freeman and Company, New York, 1981.

Stetter, K. O., G. Fiala, G. Huber, R. Huber, and A. Segerer, Hyperthermophilic microorganisms, *FEMS Microbiol. Rev.*, 75, 117-124, 1990.

Stevens, T. O., and J. P. McKinley, Lithoautotrophic microbial ecosystems in deep basalt aquifers, *Science*, 270, 450-454, 1995.

Sturchio, N. C., M. R. Antonio, L. Soderholm, S. R. Sutton, and J. C. Brannon, Tetravalent uranium in calcite, *Science*, 281, 971-973, 1998.

Teng, H. H., P. M. Dove, C. A. Orme, and J. J. De Yoreo, Thermodynamics of calcite growth: baseline for understanding biomineral formation, *Science*, 282, 724-727, 1998.

Tsuji, T., Y. Kawasaki, S. Takeshima, T. Sekiya, and S. Tanaka, A new fluorescent staining assay for visualizing living microorganisms in soil, *Appl. Environ. Microbiol.*, 61, 3415-3421, 1995.

Ullrich, S., B. Karrasch, H. -G. Hoppe, K. Jeskulke, and M. Mehrens, Toxic effects on bacterial metabolism of the redox dye 5-cyano-2,3-ditolyl tetrazolium chloride, *Appl. Environ. Microbiol.*, 85, 4587-4593, 1996.

Vasconcelos, C., J. A. McKenzie, S. Bernasconi, D. Grujic, and A. J. Tien, Microbial mediation as a possible mechanism for natural dolomite formation at low temperatures. *Nature*, 377, 220-222, 1995.

Weinbauer, M. G., and C. A. Suttle, Comparison of epifluorescence and transmission electron microscopy for counting viruses in natural marine waters, *Aquat. Microb. Ecol.*, 13, 225-232, 1997.

Weinbauer, M. G., C. Beckmann, and M. G. Höfle, Utility of green fluorescent nucleic acid dyes and aluminum oxide membrane filters for rapid epifluorescence enumeration of soil and sediment bacteria, *Appl. Environ. Microbiol.*, 64, 5000-5003, 1998.

Whitman, W.B., D. C. Coleman, and W. J. Wiebe, Prokaryotes:The unseen majority, *Proc. Natl. Acad. Sci. USA.*, 95, 6578-6583, 1998.

Land and Marine Hydrogeology
Edited by M. Taniguchi, K. Wang and T. Gamo

Stable isotopic compositions of bacterial light hydrocarbons in marginal marine sediments

F. Nakagawa[a,*,+], U. Tsunogai[a,*,+], N. Yoshida[a,*], and D.D. Adams[b]

[a]Department of Environmental Science and Technology, Interdisciplinary Graduate School of Science and Engineering, Tokyo Institute of Technology, 4259 Nagatsuta, Midori-ku, Yokohama 226-8502, Japan

[b]Environmental Science, Center for Earth and Environmental Science, State University of New York, Plattsburgh, New York, 12901, U.S.A.

Gas samples were taken from gas pockets in cores at various depths (ca. 50 ~ 600 mbsf) of organic-rich marginal marine sediments off the Southwest African Margin. These samples were analyzed for their concentrations and stable carbon isotopic compositions of ethane (C_2) and propane (C_3) to determine the isotopic characterization of microbial ethane and propane in marine sediments. In the studied sites, a microbial origin for the C_1-C_3 hydrocarbons has been suggested by the high $C_1/(C_2+C_3)$ ratios (more than 4,000) and the ^{13}C-depleted isotopic compositions of methane ($\delta^{13}C$ = -83.9 to -61.5‰PDB). The ^{13}C compositions of microbial ethane range from highly ^{13}C-depleted $\delta^{13}C$ value of -67.9 to -34.0‰ (PDB), whereas those of propane show less variation from the values of organic carbon: -38.5~-23.6‰ (PDB). Since each gas sample shows little correlation among C_1, C_2 and C_3 in their $\delta^{13}C$ values, C_1, C_2 and C_3 must have different precursor molecules. The $\delta^{13}C$ compositions of ethane are shown to be useful indicators to distinguish bacterial and thermogenic hydrocarbon gases in marine sediments.

1. INTRODUCTION

While methane is the most abundant hydrocarbon in marginal marine sediments, other light hydrocarbons accompany methane in detectable concentrations. These are probably produced through the thermogenic and/or microbial degradation processes of organic materials. Distinguishing between these two types of natural gases is important for studying various phenomena that occur within land margin sediments, such as fluid migration, carbon cycling, natural gas (including gas hydrate) generation, and microbial activities under extreme conditions.

---

[*]CREST Project, Japan Science and Technology Corporation, Kawaguchi, Saitama 332-0012, Japan
[+]Present Address: Division of Earth and Planetary Sciences, Graduate School of Science, Hokkaido University, Kita-10 Nishi 8, Kita-Ku, Sapporo 060-0810, Japan

Stable isotopic compositions of methane in anaerobic sediments have been widely used to determine the mechanism and pathways of the production and consumption of methane, because microbial methane in general is ^{13}C-depleted relative to thermogenic methane. Also, microbial methane shows wide variations in its isotopic composition as a result of the two different methanogenic pathways: acetate fermentation and CO_2 reduction (Whiticar et al., 1986; Schoell, 1988; Sugimoto and Wada, 1993). The values of $\delta^{13}C$ and δD also provide information about secondary processes after methane production, such as methane oxidation, which preferentially consume the lighter isotopes (Coleman et al., 1981; Barker and Fritz, 1981; Tyler et al., 1994; Happell et al., 1994; Alperin et al., 1988).

Although we can expect, by analogy with methane, that microbial ethane (Davis and Squires, 1954; Oremland, 1981) is also ^{13}C-depleted relative to thermogenic ethane, we know little about its isotopic composition. Oremland et al. (1988) reported ^{13}C-depleted $\delta^{13}C$ values of -55 to -35‰PDB for ethane in some anoxic lake sediments. Taylor et al. (2000) also reported ethane with $\delta^{13}C$ values of -74 to -45 ‰(PDB) in near-surface aquifers in Canada. Both of these studies concluded that the ^{13}C-depleted values are characteristic of microbial ethane. The precursor molecules and production pathways of such ^{13}C-depleted ethane, however, are still unknown (Oremland et al., 1988). Besides, no $\delta^{13}C$ data is available for microbial ethane in marine environments at present.

Measurements of hydrocarbon concentrations during the Deep Sea Drilling Project (DSDP) and the Ocean Drilling Program (ODP) showed that microbial ethane and propane are widely distributed in deep anoxic marine sediments, from just below the sulfate reduction zone to deeper areas (~ 1,300 mbsf: meter below seafloor) (Claypool and Kvenvolden, 1983), with high $C_1/(C_2+C_3)$ ratios of more than 1,000. Such high $C_1/(C_2+C_3)$ ratios, however, make it difficult to determine the isotopic compositions of C_2 and C_3. As a result, most of the isotope determinations of ethane and propane have been limited to samples with low $C_1/(C_2+C_3)$ ratios (less than 100), presumably produced by thermogenic processes. Berner and Faber (1993), for example, have determined $\delta^{13}C$ compositions of ethane and propane (Nankai accretionary prism; ODP Leg 131) only for samples from depths deeper than 600 mbsf. These deep hydrocarbons, which show less-depleted ^{13}C compositions (ethane: $\delta^{13}C$ of -35.1 to -28.3‰PDB, propane: -40.5 to -26.3‰PDB) and low $C_1/(C_2+C_3)$ ratios (less than 100), were probably produced by thermogenic processes. In this study, we determined the carbon isotopic compositions of ethane and propane in shallow (less than 50 mbsf) to mid-depth (ca. 600 mbsf) organic-rich marine sediments in which we could expect to find microbially produced ethane and propane.

2. SAMPLING LOCATIONS

Sampling was carried out off the southwestern coast of Africa (Congo, Angola, Namibia, and South Africa), during Ocean Drilling Program Leg 175 (Meyers et al., 1998). The area investigated contains one of the great upwelling regions of the world, and is characterized by organic-rich sediments, thus providing an excellent setting for studying microbial hydrocarbon generation in marine sediments. Gas samples for isotope measurements of hydrocarbons were taken from five sites, namely 1075, 1077, 1078, 1079, and 1084 (Figure 1), which can be categorized into three areas: the Lower Congo Basin (dominated by the freshwater contributions of the Congo River, seasonal coastal upwelling, and incursions of open-ocean

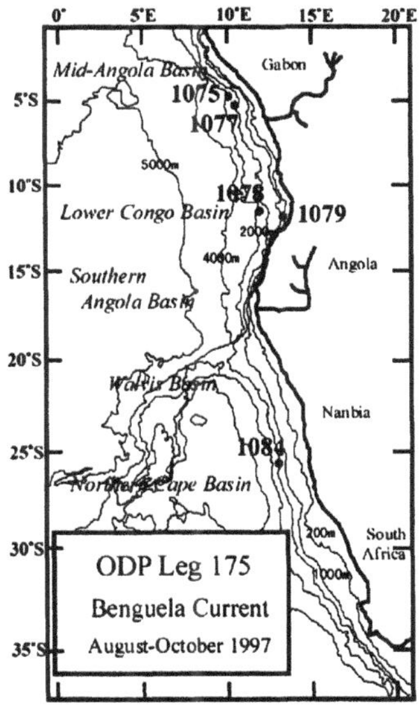

Figure 1. A topographic map showing the study sites along the south western coast of Africa.

water from the South Equatorial Countercurrent); the Angola Margin (running from the moderately productive coastal environment that contrasts with the more productive Lower Congo Basin sites to the upwelling-dominated sites south of the Angola-Benguela Front); and the Angola-Namibia Upwelling (seaward of the upwelling centers, but containing second-order signals of enhanced marine productivity associated with the upwelling signals that are transported to their respective locations by the Benguela Current and its meanders and eddies).

Sites 1075 (4°47'S, 10°05'E, 2995 m water depth) and 1077 (5°11'S, 10°26'E, 2392 m water depth) belong to the Lower Congo Basin. Sediments from these sites are composed of Pliocene-Pleistocene greenish gray diatomaceous clay and nannofossil-rich clay that contain an average of 2.5 %wt organic carbon. Sites 1078 (11°55'S, 13°24'E, 438 m water depth) and 1079 (11°56'S, 13°19'E, 749 m water depth) belong to the Angola Margin. Sediments at both sites consist of olive-gray silty clay with varying amounts of nannofossil and foraminifera, and their organic carbon concentrations are ca. 2.5 %wt, which is greater than that anticipated based on the moderate marine productivity in this sedimentary regime. Site 1084 (25°31'S, 13°02'E, 1,992 m water depth) is located on the Walvis Ridge on the northern landward edge of the Cape Basin. The sediment at this site contains similar alternations of bioturbated olive-gray to black clays rich in nannofossils and diatoms that overlie an olive-gray clayey nannofossil ooze. The sediment is notably rich in marine organic matter, and organic carbon concentrations exceed 10 %wt.

3. METHODS

Marine sediments were drilled using advanced hydraulic piston coring and the extended core barrel method. Gas pocket samples in the core liner were taken directly using a gas-tight syringe without contact with air (Meyers et al., 1998) and were stored in pre-evacuated serum bottles (30 ml) for isotope studies of methane, ethane, and propane.

Sub-samples for stable carbon and hydrogen isotope analysis of methane were processed to convert methane to CO_2 and H_2 in a vacuum line. First, after removing CO_2 and H_2O by using a liquid nitrogen temperature trap, the methane was oxidized by CuO (850°C). The CO_2 and H_2O formed by the oxidation of CH_4 were trapped at liquid nitrogen temperature, and CO_2 was subsequently separated from H_2O at dry-ice/methanol temperature. The remaining H_2O was converted to H_2 by the zinc reduction method (Coleman et al., 1982). The CH_4-derived CO_2 and H_2 were introduced into the usual dual-inlet isotope ratio mass spectrometer (Finnigan MAT252) to measure the stable carbon and hydrogen isotopic compositions of methane. Stable isotopic compositions are expressed in $\delta(‰) = ((R_{sample}/R_{standard})-1) \times 1000$, where R is $^{13}C/^{12}C$ or D/H. The $\delta^{13}C$ and δD values are reported relative to VPDB and VSMOW, respectively. The overall errors of the analyses are ±0.1‰ for $\delta^{13}C$ and ±1‰ for δD.

The stable carbon isotopic composition of trace ethane and propane were measured by a method similar to that of Tsunogai et al. (1999) using a CF-IRMS (continuous flow isotope ratio mass spectrometer) system. Sub-samples for ethane and propane measurements were passed through a pre-concentration column packed with porous glass beads immersed in liquid oxygen in a vacuum line. The pre-concentration column was then heated, and the trapped sample was flushed with ultra-pure helium and collected into a second cryofocusing trap, which was a stainless steel column (20 cm long, 1 mm i.d.) packed with Ni wires for quantitative trapping. The ultra-pure helium was obtained by passing helium (purity: >99.99995%) through two active charcoal traps held at liquid nitrogen temperature. More than 99% of the moisture and CO_2 trapped in the first pre-concentration trap were removed from the sample gas by passing it through a Drierite and Ascarite II trap during the transfer procedure to the second cryofocusing trap. The second cryofocusing trap was then heated to room temperature, and the hydrocarbons were separated using gas chromatography (column: PoraPLOT-Q, oven temperature: 30 °C to 240 °C with a heating rate of 1 to 20 °C/min, helium flow: 0.5 mL/min). The column-separated hydrocarbons were quantitatively oxidized to CO_2 by passing them through a 960 °C combustion unit (CuO/Pt catalyst) and introduced to a MAT252 isotope ratio mass spectrometer for carbon isotope ratio measurements by simultaneous monitoring of masses $^{44}CO_2$, $^{45}CO_2$, and $^{46}CO_2$.

Prior to the sample measurements, complete oxidation of C_2H_6 and C_3H_8 to CO_2 at the combustion unit was confirmed by on-line trapping of CO_2 at a liquid nitrogen trap attached after the combustion unit and by monitoring mass 28 (CO) at a mass spectrometer during the analyses of the working standard. The accuracy of the isotopic analysis was found to be better than 0.3 ‰ vs PDB by measuring an isotopic standard prepared by the U.S. National Institute of Standards and Technology (RM8560). The precision of the isotopic analysis, which is calculated from the standard deviation of the mean value by measuring the working standard in various volumes, depended mainly on the carbon size of the sample analyzed. The standard deviation of the mean was better than 0.2 ‰ for injections greater than ca. 200 pmolC. The analytical blank associated with the whole procedure of this analysis was less than the detection limit (3 pmolC). No systematic relationship between sample size and mean isotopic composition was observed in the working standard measurements, again indicating that the effects of analytical blanks were minimal.

The stable carbon isotope composition of CO_2 was also measured using a similar CF-IRMS system. First, sub-samples for CO_2 measurements were directly introduced into the line by a gas-tight syringe and purged with ultra-pure helium. The CO_2 was cryogenically trapped in a stainless steel column (20 cm long, 1 mm i.d.) packed with Ni wires and then cryofocused into a capillary trap for 60 s. After being separated from other trace contaminants on a PoraPLOT-Q analytical capillary column (25 m × 0.32 mm i.d., Chrompak Inc., NJ, USA) at a helium flow rate of 0.6 mL/min, CO_2 was introduced into a MAT 252 isotope ratio mass spectrometer.

The overall errors of the $\delta^{13}C$ analyses of CO_2 on a CF-IRMS are ±0.2‰.

4. RESULTS AND DISCUSSION

As is the case for all other marine sediments, sulfate consumption via microbial sulfate reduction occurred from the top few meters to several tens of meters at each sampling site. Methane, accompanied by trace levels of ethane and propane, began to increase just below the sulfate reduction zone, where the sulfate concentration drops below the detection limit. The depth at which methane appeared differed among the sites, mainly due to the differences in the deposition rate of organic-rich sediments (Meyers et al., 1998).

Gas samples were collected from gas pockets below the sulfate reduction zone. Concentrations of methane, ethane, propane and CO_2 with the corresponding isotope data are listed in Table 1. Methane was the most abundant gas in the gas pocket samples, ranging from 31.2% to 95.9%. The second most abundant gas was carbon dioxide (0.5% to 24.3%), while trace levels of ethane (4–222 ppmv) and propane (2–21 ppmv) were also found in the samples.

Methane showed ^{13}C-depleted $\delta^{13}C$ values of -84.0‰ to -61.9‰(PDB), which clearly demonstrates its biogenic origin. The well-known precursors of microbial methane are carbon monoxide, carbon dioxide, acetate, formate, methylamine, methanol, and dimethylsulfide (Conrad, 1989). Carbon dioxide reduction is thought to be the main formation pathway leading to methane in marine sediments (Claypool and Kaplan, 1974). A contribution from acetate fermentation pathways, however, has also been reported in some marginal areas where the sedimentation rate is relatively high. The other compounds are thought to be of minor importance in marine sediments. Carbon isotopic fractionation (α_c) between methane and carbon dioxide is defined by the following equation:

$$\alpha_C = \frac{(\delta^{13}C_{CO2} + 10^3)}{(\delta^{13}C_{CH4} + 10^3)} \tag{4.1}$$

This value is known to be useful for distinguishing the two formation pathways (CO_2 reduction and acetate fermentation) (Whiticar et al., 1986; Hornibrook et al., 2000). Carbon dioxide reduction is accompanied by a larger isotopic fractionation (greater than 1.060) than is acetate fermentation (less than 1.040). The observation of a large fractionation factor ($\alpha_{CO2\text{-}CH4}$ = 1.067–1.077) in the samples suggests that CO_2 reduction is the predominant methanogenic pathway. Excluding the data between depths of 100 mbsf and 200 mbsf at site 1084, the measured $\delta^{13}C$ values of methane tend to increase with depth in parallel with $\delta^{13}C$ values of CO_2 (Figures 2-a,b). This trend has also been reported for marine sediments in other areas (Claypool and Kaplan, 1974) and is generally understood as resulting from the production of isotopically light methane by the CO_2 reduction pathway, which results in the enrichment in ^{13}C of the remaining CO_2, in compliance with the Rayleigh model.

Table 1. Concentration and isotope composition of gases from gas pockets in marine sediments, off the Southwest African Margin (ODP Leg 175).

Site	Sample	Depth (mbsf)	CO_2 (vol.%)	CH_4 (vol.%)	C_2H_6 (ppmv)	C_3H_8 (ppmv)	$\delta^{13}C$-CO_2 (‰$_{PDB}$)	$\delta^{13}C$-CH_4 (‰$_{PDB}$)	δD-CH_4 (‰$_{SMOW}$)	$\delta^{13}C$-C_2H_6 (‰$_{PDB}$)	$\delta^{13}C$-C_3H_8 (‰$_{PDB}$)
1075	A-10H-4	82.3	3.7	88.2	142	6	-13.0	-84.0	-187	-63.4	-34.1
	A-11H-2	89.1	1.4	55.9	81	3	-12.8	-82.9	-186	-64.8	-35.0
	C-12H-1	103.7	2.6	93.9	-	-	-11.6	-80.8	-189	-	-
	C-13H-2	113.9	2.1	93.8	140	6	-9.7	-80.1	-189	-64.8	-38.5
	C-14H-1	122.6	2.9	94.0	-	-	-9.3	-79.1	-191	-	-
	C-15H-6	140.0	1.6	62.6	102	5	-8.0	-77.7	-189	-64.5	-36.5
	C-17H-6	159.1	2.5	91.0	-	-	-7.0	-76.7	-189	-	-
	C-18H-6	168.6	2.0	94.9	-	-	-6.3	-76.6	-188	-	-
	C-19H-2	170.9	2.3	93.8	166	9	-5.6	-76.0	-188	-64.5	-36.2
	C-20H-6	187.4	3.3	95.9	-	-	-5.1	-75.3	-188	-	-
	C-21H-2	190.0	1.6	52.1	-	-	-5.1	-74.8	-183	-	-
	C-22H-3	200.3	2.2	95.3	175	11	-5.2	-74.8	-188	-63.8	-33.4
1077	B-13H-3	113.4	2.1	93.1	92	11	-10.3	-80.4	-191	-67.9	-35.7
	B-14H-2	120.5	2.5	93.9	97	11	-9.5	-79.7	-190	-67.8	-36.6
1078	A-5H-3	39.9	9.8	80.9	34	9	-6.5	-76.9	-183	-34.0	-28.1
	A-6H-3	48.7	10.4	84.2	-	-	-2.20	-73.1	-186	-	-
	A-8H-3	67.6	17.1	76.9	35	10	+0.42	-70.4	-186	-35.6	-30.8
	D-8H-6	68.5	7.2	39.1	-	-	+0.32	-70.2	-180	-	-
	D-9H-4	75.3	13.1	47.9	24	7	+0.99	-69.7	-182	-37.2	-28.9
	D-10H-4	83.7	17.9	51.4	-	-	+0.35	-69.6	-183	-	-
	D-11H-2	90.3	13.7	49.2	-	-	+0.95	-69.7	-183	-	-
	D-12H-2	101.0	24.3	70.8	51	6	+0.85	-69.3	-185	-39.1	-34.9
1079	B-7H-2	57.2	3.6	88.5	138	6	-16.1	-80.8	-187	-42.5	-30.6
	B-8H-2	66.3	3.4	93.8	-	-	-7.6	-75.9	-	-	-
	A-9H-3	76.0	3.1	93.2	183	12	-4.2	-73.5	-186	-42.9	-31.5
	A-10H-3	84.1	5.6	90.5	-	-	-2.7	-71.1	-184	-	-
	A-11H-3	94.4	5.0	92.1	-	-	-0.9	-70.5	-186	-	-
	A-12H-3	104.5	4.4	94.1	200	15	-0.2	-63.8	-	-42.3	-31.6
	A-13H-4	112.8	3.6	91.0	-	-	+0.61	-67.9	-180	-	-
1080	A-4H-11	31.6	1.0	89.5	192	21	-6.0	-71.9	-190	-52.1	-30.1
	A-5H-6	37.5	0.5	94.9	-	-	-3.7	-70.2	-194	-	-
	A-6H-7	49.7	1.1	93.0	222	18	-0.6	-69.2	-190	-52.7	-31.2
1082	A-8H-3	66.7	4.3	51.3	33	3	-0.2	-72.6	-188	-55.1	-33.9
	A-10H-7	92.7	2.8	55.5	-	-	+2.71	-69.9	-183	-	-
	A-12H-3	105.4	4.0	57.9	45	5	+2.73	-69.1	-185	-54.8	-28.2
	A-45X-4	412.6	6.3	27.1	50	12	-0.6	-61.6	-170	-49.5	-26.9
1084	A-7H-6	58.4	7.5	57.4	4	2	-0.0	-71.2	-199	-48.2	-27.6
	C-13H-5	119.2	7.1	57.0	12	2	+4.09	-67.0	-198	-58.8	-33.5
	C-15H-5	138.0	11.4	58.0	-	-	+3.91	-66.1	-198	-	-
	C-16H-4	146.5	7.3	31.2	-	-	+3.49	-65.4	-201	-	-
	C-17H-5	156.8	9.6	58.3	-	-	+4.31	-63.6	-209	-	-
	C-18H-2	162.3	8.5	56.9	8	3	+3.28	-64.1	-198	-55.9	-30.6
	C-19H-6	175.3	6.2	65.4	-	-	+2.62	-63.6	-204	-	-
	C-20H-2	181.4	9.7	58.0	10	4	+2.86	-62.9	-199	-59.4	-33.7
	C-22H-6	205.4	7.3	84.9	58	2	+1.92	-61.9	-208	-59.6	-23.6
	A-40X-5	362.1	7.8	55.5	65	6	+1.77	-64.3	-187	-56.0	-25.3
	A-41X-6	372.0	6.0	46.0	-	-	+2.34	-64.5	-	-	-
	A-42X-3	377.9	21.5	42.5	-	-	+1.85	-64.2	-187	-	-
	A-47X-3	424.9	16.7	47.5	78	9	+0.63	-64.9	-187	-54.9	-27.4
	A-62X-4	570.7	10.8	58.6	119	20	-3.3	-66.0	-192	-52.2	-28.1
	A-62X-6	574.3	7.8	63.4	-	-	-2.9	-66.5	-184	-	-
	A-65X-4	601.4	5.0	64.1	90	17	-4.5	-66.7	-190	-50.8	-28.6

The hydrogen isotopic composition of methane was uniform with depth (Figure 2-c), with a mean value of -190±7‰ (n=45). These results were similar to those reported in other marine sediments (Whiticar et al., 1986) where methane was produced by microbial processes and was found to be enriched in deuterium by comparison with methane in freshwater environments,

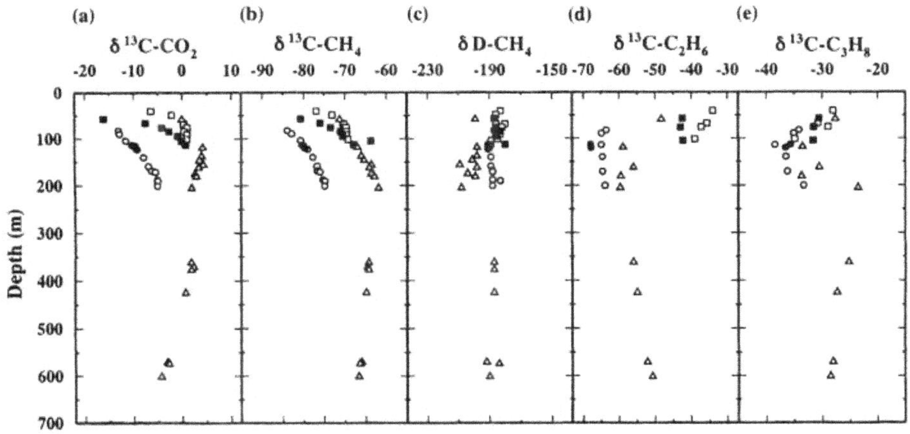

Figure 2. Isotopic compositions of gases from gas pockets in marine sediments, off the Southwest African Margin (ODP Leg 175). (a) δ^{13}C-carbon dioxide, (b) δ^{13}C-methane, (c) δD-methane, (d) δ^{13}C-ethane, and (e) δ^{13}C-propane. Legends are expressed as ○: site 1075, ●: site 1077, □: site 1078, ■: site 1079, and △: site 1084.

such as wetland and rice paddies. Although no data are available for the δD of pore water at this site, the fractionation factor between methane and water, which is defined by following equation:

$$\varepsilon_D = \frac{(\delta D_{CH4} - \delta D_{H2O})}{(\delta D_{H2O} + 10^3)} \times 10^3 \qquad (4.2)$$

can be estimated to be 190±7‰, by assuming the δD of water to be 0‰, the δD value of ocean water. This fractionation factor is significantly lower (i.e., more than 100‰ lower) than that obtained experimentally by incubating freshwater soil (ε_D = 317±20‰; Sugimoto and Wada, 1993). The relatively small hydrogen isotopic fractionation between water and methane might be due to a lower hydrogen gas concentration during CO_2 reduction, suggesting that the methane precursor (H_2) in marine sediments is depleted at each site due to the slow degradation rates of organic matter (Burke, 1993).

Gases sampled between the depths of 100 mbsf to 200 mbsf at site 1084 showed some different profiles in the stable isotopic composition of methane and CO_2 then were shown at the other sites. An inverse vertical trend was observed between δ^{13}C of CO_2 and that of methane, where the former became lighter while the latter became heavier with increasing depth. One possible reason for this carbon isotopic trend is anaerobic methane oxidation, which preferentially consumes light methane to result in the production of light CO_2 and remaining heavy methane. However, the absence of deuterium-enrichment in methane suggests that methane oxidation in this zone is negligible. The other possible explanation for this trend is a difference in the methanogenic pathways at this site. This site is characterized by an extremely high organic carbon content in the marine sediment (about 10%). This high carbon content will lead to an enhancement of the proportion of methane produced via the acetate fermentation pathway, which produces both lighter δ^{13}C in CO_2 relative to the δ^{13}C in CO_2 observed at the top of this zone (+4.1‰), and δ^{13}C-enriched methane (more than -40‰). The δD value in this zone was somewhat depleted in deuterium (mean value: -202±5‰, n=8) relative to the value at other sites (mean value: -187±4‰, n=37). The greater abundance of organic carbon may elevate the H_2 concentration and also the concentration of CH_4 derived through acetate fermentation, both of which result in lower δD values in methane.

The $C_1/(C_2+C_3)$ ratio is also a useful index that can be used to distinguish between the microbial (ratio of more than 1,000) and thermogenic (less than 100) origins of hydrocarbons. The $C_1/(C_2+C_3)$ ratios of all the hydrocarbons (C_1–C_3) at these sites were in the range of 4,213 ~95,733, indicating a microbial origin for the hydrocarbons C_1–C_3 (Davis and Squires, 1954; Oremland, 1981).

The ^{13}C compositions of the biogenic ethane show wide variations, ranging from -67.9 to -34.0‰ (Table 1). The $\delta^{13}C$ values of ethane are highly depleted in ^{13}C relative to those sampled from most natural gas or oil fields (see below) and those from surface anaerobic sediments of some lakes. It is impossible to explain the lower $\delta^{13}C$ values as a result of thermogenic processes unless we assume that the $\delta^{13}C$ of the precursor molecule has highly ^{13}C -depleted $\delta^{13}C$ of -40‰ or less (Berner and Faber, 1996). The minimum value of -67.9‰ represents the $\delta^{13}C$ value of biogenic ethane.

Previous isotopic studies of ethane and propane have been concentrated in natural gas-oil fields, and have focused on the assessment of the origin and stages of the generation of natural gas. Most ethane and propane in natural gas show $\delta^{13}C$ values of > -35‰ and > -30‰, respectively, for propane, indicating a thermogenic origin. In some natural gas or oil fields, however, ^{13}C-depleted ethane has been reported, with values of -62 to -45‰ being reported in Tertiary subalpine basins (Schoell, 1984), -58 to -20‰ in the Sacramento basin (Jenden and Kaplan, 1989), –58 to -37‰ in the northern Apennine (Mattavelli et al., 1992), and –63 to –32 ‰ in the western Canadian sedimentary basin (Rowe and Muehlenbachs, 1999). The above authors have concluded that these must be low-temperature thermogenic ethane, since natural gases in these fields contain ethane concentrations well over 0.1 %. Since methane is selectively reduced through microbial oxidation (Whiticar and Faber, 1985), however, the methane / ethane ratio must be understood as an unreliable indicator for thermogenic hydrocarbons. As revealed in this study, biogenic ethane is ubiquitously depleted in ^{13}C. Small contributions of such ^{13}C-depleted biogenic ethane could be also responsible for the ^{13}C-depleted ethane even in these natural gas-oil fields.

Although the mechanisms of the microbial formation of ethane are far from understood, certain methanogenic bacteria are known to be responsible for the ethane formation processes in anaerobic sediments (Oremland, 1981). If ethane is produced from a common source as a by-product of methane by methanogenic bacteria, the trend of $\delta^{13}C$-ethane should be similar to that of methane. The $\delta^{13}C$ profile of ethane, however, showed no relationship with that of methane, showing only minor changes with depth (Figure 2-d). The ethane is produced by a different formation process using different substrates from those of methane. Another striking feature of the $\delta^{13}C$ distribution of ethane is that the variations among the sites are greater than the variation with depth in each profile. The $\delta^{13}C$ values of unknown precursors of ethane seem to be different at different sites, a finding that may be caused by differences in the type of substrate or in the pool size of the substrate. At present there are no experimental data that can explain the highly ^{13}C-depleted ethane observed in anaerobic sediments (Oremland et al., 1988). Oremland et al. (1988), for example, have determined the stable carbon isotopic fractionation associated with ethane formation from diethylsulfides to be -4.6‰, and that from ethanethiol to be -6.5‰ by sediment slurry incubations. It is difficult to explain the observation of

^{13}C-depleted ethane based on such small fractionations. Further laboratory culture studies are necessary.

Figure 2-e also shows the $\delta^{13}C$ profile of propane ranging from -38.5 to -23.6‰, which is relatively close to the profile of organic carbon. As is the case for ethane, trace levels of propane in shallow anaerobic sediments might also be produced via a microbial process, although the mechanisms are not yet known. Considering that propane is a three-carbon molecule, one can expect that its carbon isotopic fractionation during formation will be roughly one-third that of methane and two-thirds that of ethane, if these hydrocarbons are produced from a common source. In other words, the $\delta^{13}C$ of microbial propane must be less negative than either ethane or methane. The $\delta^{13}C$ value of propane obtained in marine sediments seems to be reasonable, because of a ^{13}C-enriched $\delta^{13}C$ value relative to that observed for ethane and methane. The trend of the $\delta^{13}C$ profile of propane, however, differed from those of methane and ethane. The formation processes of propane, therefore, may also differ from that of methane and even that of ethane. The formation process of propane should also be elucidated in future studies.

ACKNOWLEDGEMENTS

The samples used in this study were provided by the Ocean Drilling Program (ODP). ODP is sponsored by the U.S. National Science Foundation (NSF) and participating countries under the management of Joint Oceanographic Institutions (JOI), Inc. This research is partly funded by the Ministry of Education, Science & Technology through the Special Coordination Fund "Archaean Park" project. We thank anonymous reviewers for their constructive comments.

REFERENCES

Alperin, M.J., Reeburgh, W.S. and Whiticar, M.J., 1988. Carbon and hydrogen isotope fractionation resulting from anaerobic methane oxidation. Global Biogeochem. Cycles 2, 279-288.

Barker, J.F. and Fritz, P., 1981. Carbon isotopic fractionation during microbial methane oxidation. Nature 293, 289-291.

Berner, U. and Faber, E., 1993. Light hydrocarbons in sediments of the Nankai accretionary prism (Leg 131, Site 808). Proc. ODP, Sci. Results, 131, College Station, TX (Ocean Drilling Program), pp. 185-195.

Berner, U. and Faber, E., 1996. Empirical Carbon Isotope/Maturity Relationships for Gases from Algal Kerogens and Terrigenous Organic-Matter, Based on Dry, Open-System Pyrolysis. Org. Geochem. 24(10-11), 947-955.

Burke, Jr., R.A., 1993. Possible influence of hydrogen concentration on microbial methane stable hydrogen isotopic composition. Chemosphere 26, 55-67.

Claypool, G. and Kvenvolden, K., 1983. Methane and other hydrocarbon gases in marine sediments. Ann. Rev. Earth Planet. Sci. 11, 299-327.

Claypool, G.E. and Kaplan, I.R., 1974. The origin and distribution of methane in marine sediments. In: Kaplan, I.R. (Ed.), Natural gases in marine sediments. Plenum, New York, pp. 99-139.

Coleman, D.D., Risatti, J.B. and Schoell, M., 1981. Fractionation of carbon and hydrogen isotopes by methane-oxidizing bacteria. Geochim. Cosmochim. Acta 45, 1033-1037.

Coleman, M.L., Shepard, T.J., Durham, J.J., Rouse, J.E. and Moore, G.R., 1982. Reduction of water with zinc for hydrogen isotope analysis. Anal. Chem. 54, 993-995.

Conrad, R., 1989. Control of methane production in terrestrial ecosystems, In: Andreae, M.O. and Schimel, D.S. (Eds.), Exchange of trace gases between terrestrial ecosystems and the atmosphere. John Wiley & Sons, Inc, New York, pp. 39-58.

Davis, J.B. and Squires, R.M., 1954. Detection of microbially produced gaseous hydrocarbons other than methane. Science 119, 381-382.

Happel, J.D., Chanton, J.P. and Showers, W.S., 1994. The influence of methane oxidation on the stable isotopic composition of methane emitted from Florida swamp forest. Geochim. Cosmochim. Acta 58, 4377-4388.

Hornibrook, E.R.C., Longstaffe, F.J. and Fyfe, W.S., 2000. Evolution of stable carbon isotope compositions for methane and carbon dioxide in freshwater wetlands and other anaerobic environments. Geochim. Cosmochim. Acta 64, 1013-1027.

Jenden, P.D., and Kaplan, I.R., 1989. Origin of natural gases in Sacramento Basin California. AAPG Bull. 73, 431-453.

Mattavelli, L., Ricchiuto, T. and Martinenghi, C., 1992. Deep isotopic light methane in northern Italy. In: Vially, R. (Ed.), Bacterial Gas. Editions Technip, Paris, pp. 121-132.

Meyers, P.A. and shipboard scientific party, 1998. Microbial gases in sediments from the southwest African margin. Proc. ODP, Initial. Rep., 175, College Station, TX (Ocean Drilling Program), pp. 555-560.

Oremland, R.S., 1981. Micobial formation of ethane in anoxic estuarine sediments. Appl. Environ. Microbiol. 42, 122-129.

Oremland, R.S., Whiticar, M.J., Strohmaier, F.E. and Kiene, R.P., 1988. Bacterial ethane formation from reduced, ethylated sulfur compounds in anoxic sediments. Geochim. Cosmochim. Acta 52, 1895-1904.

Rowe, D. and Muehlenbachs, A., 1999. Low-temperature thermal generation of hydrocarbon gases in shallow shales. Nature 398, 61-63.

Schoell, M., 1988. Multiple origin of methane in the earth. Chem. Geol. 71, 1-10.

Schoell, M., 1984. Wasserstoff- und Kohlenstoffisotope in organischen Substanzen, Erdolen und Erdgasen. Geologisches Jahrbuch 67, 164.

Sugimoto, A. and Wada, E., 1993. Carbon isotopic composition of bacterial methane in a soil incubation experiment: Contribution of acetate and CO_2/H_2. Geochim. Cosmochim. Acta 57, 4015-4027.

Taylor, S.W., Sherwood, B. and Wassenaar, L.I., 2000. Bacteriogenic ethane in near-surface aquifers: Implications for leaking hydrocarbon well bores. Environ. Sci. Tech. 34, 4727-4732.

Tsunogai, U., Yoshida, N. and Gamo, T., 1999. Carbon isotopic compositions of C_2-C_5 hydrocarbons and methyl chloride in urban, coastal and maritime atmosphere over western-North Pacific. J. Geophys. Res. 104, 16,033-16,039.

Tyler, S.C., Crill, P.M. and Brailsford, G.W., 1994. $^{13}C/^{12}C$ fractionation of methane during oxidation in a temperate forested soil. Geochim. Cosmochim. Acta 58, 1625-1633.

Whiticar, M.J. and Faber, E., 1985. Methane oxidation in sediment and water column environments-isotope evidence. Org. Geochem. 10, 759-768.

Whiticar, M.J., Faber, E. and Schoell, M., 1986. Biogenic methane formation in marine and freshwater environments: CO_2 reduction vs. acetate fermentation-isotope evidence. Geochim. Cosmochim. Acta 50, 693-709.

Land and Marine Hydrogeology
Edited by M. Taniguchi, K. Wang and T. Gamo

Submarine hydrothermal activity in coastal zones

Toshitaka Gamo[a] and Geoffrey P. Glasby[b]

[a]Division of Earth and Planetary Sciences, Graduate School of Science, Hokkaido University, N10 W8, Sapporo 060-0810, Japan

[b]Laboratory for Earthquake Chemistry, Graduate School of Science, University of Tokyo, 7-3-1 Hongo, Bunkyo-ku, Tokyo 113-0033, Japan

Coastal hydrothermal acitivity has been observed at several sites in the Mediterranean Sea and western Pacific and is almost always associated with arc volcanism at subduction zones. While deep-sea hydrothermal fluids are mainly derived from the circulation of seawater beneath the seafloor, coastal hydrothermal fluids are derived from a complex mixture of seawater, meteoric water (groundwater) and magmatic fluids. In addition, tidal forcing, sea level change and earthquake activity may significantly affect the rates of fluid venting and dispersion of hydrothermal plumes. The chemical composition of coastal hydrothermal fluids is therefore variable because it depends not only on fluid-rock interaction at high temperatures but also on the rate of subduction of the slab material at the convergent plate margin and the decomposition of organic matter within the coastal sediments through which the fluids pass. Biological communities in the coastal hydrothermal system have complicated characteristics owing to the simultaneous occurrence of chemosynthetic and photosynthetic primary production.

1. INTRODUCTION

Although most studies of submarine hydrothermal activity have focussed on deep-sea systems, volcanism in shallow coastal zones, on the flanks of volcanic islands and on the tops of seamounts may also cause hydrothermal circulation similar to that found at mid-oceanic ridges and in back-arc basins. One of the remarkable characteristics of coastal hydrothermal activity is its chemical complexity, since the coastal fluids may be derived from meteoric water (groundwater), seawater and magmatic fluid in variable amounts, whereas deep-sea hydrothermal fluids are derived exclusively from seawater except in special cases where magmatic fluids are involved (de Ronde, 1995; Gamo et al., 1997). In addition, shallow-water hydrothermal systems are different from those of the deep sea in that the lower hydrostatic pressure facilitates boiling (phase separation) of the ascending fluid resulting in higher contents of magmatic volatiles.

Submarine hydrothermal activity in coastal zones may significantly affect biogeochemical cycles across the land-ocean boundary and within poorly-flushed coastal marine environments. Although submarine hydrothermal activity is known to play an important role in the geochemical cycle between the ocean and the lithosphere (e.g. Thompson, 1983; Von Damm, 1990; 1995; Kadko et al. 1995; Elderfield and Schulz, 1996), little is known about the magnitude of geochemical fluxes in coastal systems (Dando et al., 2000; Moore and Brewer, 2001).

Submarine hydrothermal activity has been observed and investigated at many coastal sites. It is usually associated with volcanic activity on land, predominantly arc but occasionally hot spot volcanism. The Smithsonian National Museum of Natural History has published a map showing the locations of active volcanoes on the surface of the earth together with a description of each volcano (Anon., 2001a). From these descriptions, the following volcanoes appear to be associated with shallow submarine hydrothermal activity and suitable for further investigation: Vulcano and Stromboli (Aeolian Islands), Milos, Santorini, Yali and Kos (Aegean Arc), Piton de la Fournaise and McDonald Island (Indian Ocean), Hunga Tonga-Hunga Ha'apai, Metis Shoal, Home Reef, Fonualei and Curacoa (Tonga), Vailulu'u (Samaoa), White Island, Whale Island and Calypso Seamount (New Zealand), Curtis Island, Raoul Island and Monowai Seamount (Kermadec Islands), Rabaul (New Britain), Kavachi (Solomon Islands), Tinakula (Santa Cruz Island), Gaua, Kuwae and Eastern Gemini Seamount (Vanuatu), Barren Island (Andaman Islands), Krakatau, Iliwerung and Banua Wuhu (Indonesia), Dikidas (Philippines), Izu-Tobu, Ryukyu Islands, Izu Islands (Japan), Ushishur (Kuril Islands), Mariana Islands, Loihi Seamount (Hawaii), Teahitia and Mehitia (French Polynesia), Socorro Island (Mexico), Kick-'em Jenny (West Indies), Azores, Kolbeinsey Ridge (Iceland) and Deception Island (Antarctica).

Examples of coastal zones where studies of submarine hydrothermal activity have been carried out include the areas around the island-arc volcanoes of the northern Mediterranean Sea (Sedwick and Stüben, 1996; Stüben et al., 1996; Dando et al., 1995a,b, 1999, 2000; Stüben and Glasby, 1999), Bay of Plenty, New Zealand (Sarano et al., 1989; Pantin and Wright, 1994), island-arc and forearc volcanoes in Papua New Guinea (Pichler et al., 1999; Tarasov et al., 1999), Mahengetang island, Indonesia (Heikoop et al., 1996), Kagoshima Bay southwest of Japan (Kagoshima Prefecture, 1977; Horibe et al., 1980; Craig and Horibe, 1994; Yamanaka et al., 1999; 2001), off the Japanese coast (Sakai and Matsubaya, 1974; 1977), Kuril Islands in Russia (Tarasov et al., 1990) and northern Baja California, U.S.A. (Vidal et al., 1978, 1981).

In this chapter, we have selected several representative sites in the Mediterranean Sea and western Pacific Ocean for review, mainly from a biogeochemical standpoint. Figure 1 shows the locations of these sites. For the purposes of this chapter, we need to define the coastal zone. Holligan and Booijs (1993) defined the coastal zone as extending from the coastal plains to the outer edge of the continental shelves, approximately matching the region that has been flooded and exposed during the sea level fluctuations of the late Quaternary period. According to Crossland and Kremer (2002), the coastal zone encompasses river basins and catchments, estuaries and coastal seas and extends to the continental shelf --- the domain surrounding the land-sea interface extending to the landward and seaward limits of marine and terrestrial influences. Although there is no single definition for the coastal zone, there is general acceptance that the definition needs to vary according to the type of problem being addressed and the objectives of management. A common rule-of-thumb is to include the landward area to 100 km from the coast (WRI, 2000). In LOICZ (Land-Ocean Interactions in the Coastal Zone), the coastal zone is nominally considered to extend from 200m elevation to 200m depth (Pernetta and Milliman, 1995).

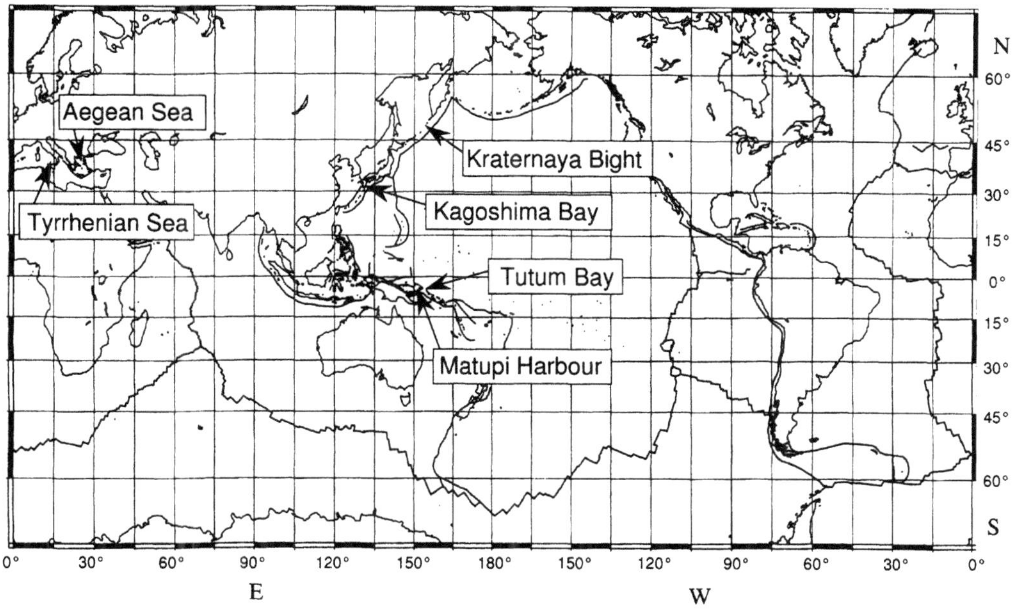

Figure 1. A global tectonic map showing the locations of the hydrothermal systems described in the text.

2. EXAMPLES OF SUBMARINE HYDROTHERMAL ACTIVITY IN COASTAL ZONES

2.1. Mediterranean Sea (Tyrrhenian Sea and Aegean Sea)

Dando et al. (1999) have described hydrothermalism in the Mediterranean Sea in detail referring to many previous works. Volcanic and hydrothermal activity in the Mediterranean occurs as a result of subduction of the African plate beneath the Eurasian plate at a rate of 10-30 mm yr^{-1}. The subduction is characterized by complex tectonics with a number of microplates. There are active arc volcanoes and back-arc basins north of the plate boundary associated with hydrothermal circulation systems. The presently active volcanoes are Etna, Vulcano, Stromboli and Vesuvius in Italy and Santorini and Nisyros in Greece.

Most of the presently known hydrothermal sites are located in shallow coastal areas of <200 m depth, in the Tyrrhenian Sea (Italy) and the Aegean Sea (Greece) (see Fig. 1). They usually occur as a gasohydrothermal vent type, probably involving degassing from the subducted slab and mantle and the decomposition of carbonates in marine sediments. The major component of the vented gas is CO_2 with SO_2, H_2S, CH_4 and H_2 as minor components. Stable isotope studies have confirmed a magmatic contribution to the released gases in the Aegean Sea (Nagao et al., 1991; Botz et al., 1996). Dando et al. (2000) provisionally estimated the total flux of CO_2 and CH_4 as free gas from the Milos submarine hydrothermal system to be 0.2-0.8 x 10^{11} moles yr^{-1} and 0.2-0.8 x 10^{9} moles yr^{-1}, respectively. These values correspond to approximately 10% of the global fluxes estimated for mid-ocean ridges (Welhan and Craig, 1983; Gerlach, 1989; Marty et al. 1989; Williams et al. 1992) indicating a considerable influence of the Mediterranean coastal hydrothermal activity on global geochemical cycles (Kerrick et al. 1995). A similar influence may be expected at other convergent plate margins.

The venting fluids show a range of salinities because of phase separation resulting from fluid boiling at low pressures as well as the introduction of meteoric water (or groundwater) at some sites. The maximum temperatures for the venting fluids from the Tyrrhenian Sea are in the range of 24-103°C, and those from the Aegean Sea of 33-119°C. The lowest pH is 5.2 in both seas. Enrichment of heavy metals such as Mn and Fe is observed in surficial sediments probably as a result of acid leaching of the metals from the underlying rocks. Massive sulphides, pyrite, haematite, sphalerite, galena and barite have also been observed at some sites (Dando et al., 1999).

Long-term records of temperature, salinity, dissolved oxygen and pH at the submarine warm spring site in the limestone cavern of Grotta Azazurra, Tyrrhenian Sea, reflect tidal forcing, with venting of the relatively warm, low-salinity, low-oxygen, acidic fluids occurring at low tides (Stüben et al., 1996). Dando et al. (1999) have demonstrated an inverse relationship between vent outlet temperature and water level for two vents off Methana in the Aegean Sea. Dando et al. (1995a) also showed that earthquakes increase hydrothermal venting and nutrient inputs into the Aegean Sea. A seismic network in the area of Milos revealed that the absence of micro-earthquakes is generally related to absence of hydrothermal signals and that a peak value of bottom water pressure coincides with a high concentration of micro-earthquakes and hydrothermal signals (Dando et al., 2000).

Although no extant vent-specific fauna have been described from Mediterranean sites, a high diversity of bacteria, diatom, and epifauna has been reported and the vent sites are areas of settlement for exotic thermophilic species. Large numbers of prokaryotes, especially hyperthermophilic crenarchaeota, have been isolated from the Mediterranean hydrothermal vents (Dando et al., 1999).

2.2. Pacific Ocean

2.2.1. Matupi Harbour and Tutum Bay (Papua New Guinea)

Matupi harbour is a relatively sheltered basin with a maximum depth of 65 m and an area of ~4 km^2 located at Rabaul, New Britain (see Fig. 1). It is surrounded by several active volcanoes. Tarasov et al. (1999) identified three groups of littoral and sublittoral thermal springs in the harbour. Bottom surveys by SCUBA to a depth of 40 m identified gasohydrothermal fields where gas bubbles are released from the bottom with temperatures of 40 to 90°C. The hydrothermal fluids are derived mainly from seawater but are significantly enriched in CO_2, CH_4, H_2, Fe, Mn, Zn, Pb, Cd, Si, H_2S and NH_4^+ and have a minimum pH of 3.1.

Unique characteristics of the surface waters (0-3 m layer) were observed in the harbour as a result of the inflow of hot hydrothermal fluids which are rich in nutrients and reduced compounds and therefore stimulate primary production by both photosynthesis and chemosynthesis. Bacterial mats, filamentous colourless sulphur bacteria, filamentous cyanobacteria and various thermophilic bacteria flourish in the venting areas. Rich benthic communities, epifauna dominated by corals and sponges and infauna by nematodes, are observed in areas adjacent to the hydrothermal vents but are sparse at the vents.

Pichler et al. (1999) and Pichler and Veizer (1999) reported another coastal hydrothermal site which is located in a fore-arc Ambitle Island, east of New Ireland. Submarine hydrothermal venting occurs at Tutum Bay (see Fig. 1) in shallow (5-10 m) water on the west coast of the island where the inner shelf contains a patchy distribution of coral-algal reefs. There are two types of venting: (1) focussed discharge of a clear, two-phase fluid with a temperature of 89-98°C and pH 6.2-6.7 from discrete orifices (0.1-0.15 m in diameter), and (2) dispersed or diffuse discharge of streams of gas bubbles. The gas is mainly CO_2 (93-98%) with minor amounts of N_2, O_2, CH_4 and He.

The chemical characteristics of the fluids at Tutum Bay reflect the following sequence below the coastal zone (Pichler et al., 1999). Phase separation in the deep reservoir produces a rising vapour phase which mixes with cooler groundwater to form a low pH, CO_2-rich water with a temperature of 150-160°C. This fluid is highly reactive and leaches pH- and redox-sensitive elements such as Fe, Mn, Ca and Sr from the host rock in the shallow reservoir. After lateral movement of the fluid toward the margin of the hydrothermal system, entrainment of minor amounts of groundwater or seawater occurs during its final ascent to venting. The hydrothermal fluids are depleted in the major components of seawater (Cl^-, Br^-, SO_4^{2-}, Na^+, K^+, Ca^{2+}, Mg^{2+}, Sr^{2+}) but enriched in HCO_3^-, B, Si, Li, Mn, Fe, Rb, Cs, Sb, Tl and As.

2.2.2. Kagoshima Bay (southwest Japan)

Kagoshima Bay is a large submerged caldera in the southern part of Kyushu Island, Japan, where many calderas have formed since the early Quaternary (see Fig. 1). Mt. Sakura-jima, which separates Kagoshima Bay into its northern and southern parts, is still actively erupting. Submersible observations showed that gas bubbles are emitted from a caldera (Wakamiko Caldera) at a depth of ~200 m in the northeastern part of the bay and that the temperature inside some vents is more than 200°C (Kagoshima Prefecture, 1978). The main component of the gas is CO_2 with minor amounts of CH_4, N_2 and H_2S (Ossaka et al., 1992), most of which is dissolved in the bottom seawater immediately after discharge with only a small amount of residual gas reaching the surface. The northern part of the bay is stagnant except in winter (Kamada et al., 1977) and the pH of the bottom water in the caldera at summer time is about 6 due to the dissolution of CO_2.

Horibe et al. (1980) applied a one dimensional diffusion model to the vertical profiles of ΣCO_2 and excess ^{222}Rn just above the caldera in September and November 1977 and estimated the vertical eddy diffusivity just above the bottom to be 88 ± 13 cm^2s^{-1} and the CO_2 flux from seafloor $1.01 \pm 0.17 \times 10^{-8}$ mol $cm^{-2}s^{-1}$. Craig and Horibe (1994) sampled the volcanic gases dissolved in bottom seawater above the caldera and showed that the helium isotope ratio ($^3He/^4He$) is $6.1R_A$ where R_A is the $^3He/^4He$ ratio of atmospheric helium which is similar to the ratios determined in Japanese land volcanoes such as Hakone and Usu volcanoes. They also found that the $CH_4/^3He$ ratio is ~1,000 times greater than the MORB (Mid Oceanic Ridge Basalt) ratio suggesting the contribution of a large quantity of thermogenic CH_4 from sedimentary organic matter as observed in the Guaymas Basin (Welhan, 1988).

Yamanaka et al. (1999a; 1999b; 2000) reported petroleum generation in hydrothermally-altered recent sediment of the caldera. In order to elucidate the chemical conditions for petroleum formation, the chemical characteristics of the venting water were also determined. Yamanaka et al. (1999a) showed that the Cl^- concentration in the venting fluids is significantly lower than in ambient seawater suggesting phase separation and/or the introduction of groundwater.

Hashimoto et al. (1993) found tube-worm communities at the bottom of a small knoll (depth: 82 m) east of the Wakamiko Caldera. This is not only the shallowest occurrence ever recorded but also the first discovery of euphotic zone tube worms based on chemosynthesis.

On the basis of sulphur isotope studies, Yamanaka et al. (2001) suggested that the life style of the tube worm is comparable with that of the cold seep communities which depend on sulphide derived from bacterial sulphate reduction using fumarolic CH_4 as an electron donor.

2.2.3. Kraternaya Bight (Kurile Islands, Russia)

Tarasov et al. (1990) reported extensively on the results on the 1985-1987 experiments on the shallow-water hydrothermal fields and areas of seeping hydrothermal fluids associated with bacterial mats and rich bottom fauna (Tarasov et al., 1990 and references herein). Kraternaya Bight is a circular submerged crater of the Ushishir Volcano with a diameter of ~1 km and maximum depth of 63 m which is thought to have formed about 10,000 years ago (see Fig. 1). A narrow strait connects the bight with the open Pacific Ocean. Terrestrial volcanic activity characterized by hot springs (92-96°C) occurs in the intertidal zone in the southeastern part of the bight. These volcanic waters are acidic (pH 2-3.5) and contain high concentrations of H_2S (up to 340 μM), other reduced sulphur compounds, ammonia, silica (up to 3 mM), phosphorus, and 2-3 orders of magnitude higher concentrations of heavy metals such as Mn, Fe, Zn, Cu, Cd, Ni, and Cr than seawater.

Around the perimeter of the bight are numerous gas vents and areas where hydrothermal fluid with a temperature of 10 to 34°C is seeping through the seafloor at depths of 0 to 22 m. The major components of the vent gases are CO_2 (54-66%) and N_2 (27-42%). The effects of the hydrothermal fluid inflow are reflected as distinctive peaks of ΣCO_2, pH, and H_2S in the water column. Bacterial, algobacterial and diatom mats occur on sloping areas of the seafloor where hydrothermal fluids are seeping in as well as near gasohydrothermal vents. These mats are thought to serve as biogeochemical filters which transform the hydrothermal trace elements and reduced compounds into sedimentary organic matter. In particular, the 5 mm thick algobacterial mats have a multi-layered structure which includes many microorganisms (different kinds of sulphur oxidizing bacteria, thermophylic archaebacteria, etc.) and are characterized by extremely high rates of CO_2 fixation and organic matter production (up to 33.4 gC m^{-2} d^{-1}). The influence of both chemosynthetic and photosynthetic pathways may result in complicated metabolic and energy cycles. The bottom macrofauna have a high population density and biomass (up to 10 kg m^{-2}).

Seawater exchange between the bight and the open sea is controlled by tides. Seawater inflow into the intermediate (4-20 m) layer of the bight occurs during the rising tide; the difference in water level between high and low tides is up to 1 m. The water flowing out from the bight has a higher temperature and lower salinity than the inflowing water and is rich in suspended material and trace metal contents as a result of the inflow of hydrothermal fluids. Dissolved metals such as Mn and Fe from gasothermal waters precipitate on suspended particles of organic origin when the pH increases.

3. SAMPLING TECHNIQUES

SCUBA diving is commonly used in bottom surveys of coastal hydrothermal sites where gasothermal fields are located by visual observation of gas bubbles released from the bottom and by measurement of water temperature. Gas, bottom sediments, hydrothermal fluids, microorganisms and other biological materials can be sampled to a depth of ~40 m.

Vent samples are collected either by inserting a Teflon tube as far as possible into the vent orifice or covering the vent orifice with a Teflon funnel (Pichler et al., 1999). Fluid samples are stored in acid-cleaned polyethylene flow-through sample bottles or syringes. Gas samples are taken in 0.5 or 1 litre air-tight bottles for analysis by gas chromatography. Waters just above the bottom and sediment samples are collected in plastic/glass bottles using an underwater electric pump. Sediments and fauna are taken using a corer.

Water samples are collected below the maximum depth of SCUBA diving using a pump sampling system or hydrocast apparatus such as a CTD-multi sampling system which are lowered from a research vessel. A submersible or ROV may be used if available.

4. COMMON CHARACTERISTICS OF COASTAL HYDROTHERMAL ACTIVITY

4.1. Hydrological setting

Figure 2 shows the patterns of fluid flow in a wide variety of settings. It is seen that fluid flow proximal to the continents is driven either by direct inflow of fluid from land, by the slow compaction of sediments imported into the subduction system or by dehydration of the subducting slab (Anon., 2001b). The last two cases generally result in the well-known cold seepage observed at deep-sea convergent plate margins (Herzig and Hannington, 2000) such as at the Nankai Trough off Japan (Moore, Taira, Klaus et al., 2001) and are not relevant to this study. We therefore focus our attention on the direct inflow of fluids from land.

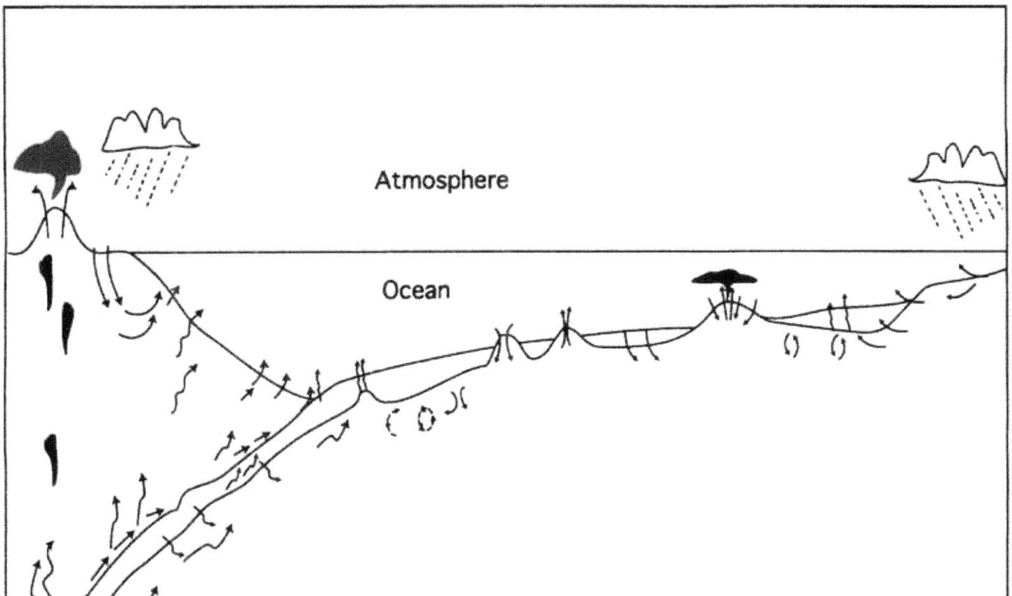

Figure 2. Schematic cross section showing possible patterns of fluid flow (arrows) through the subseafloor in a variety of hydrological settings (revised from Anon. (2001b)).

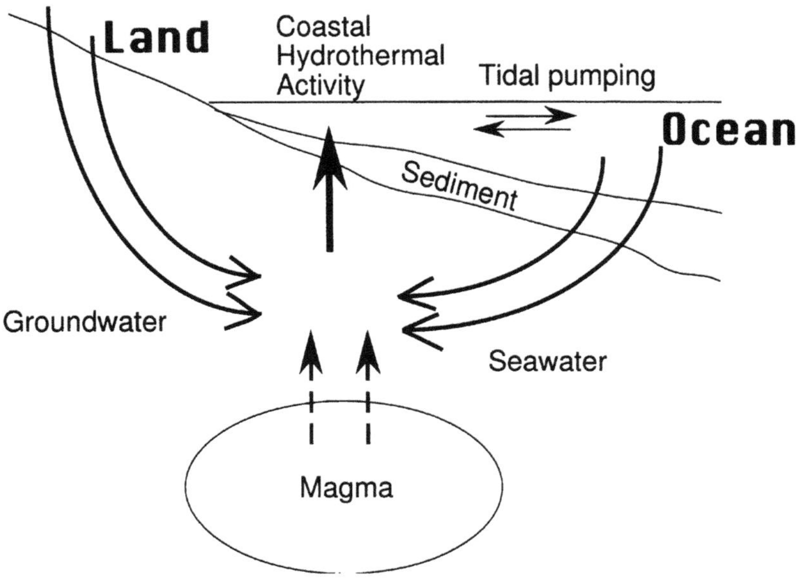

Figure 3. Schematic cross section illustrating the main inputs of fluids and gases into a coastal hydrothermal system.

A key question is how far fluids from land are able to migrate within geological strata offshore and what sort of pressure head is required to drive this migration. Although no specific answer can be given to this question, it is known that large-scale lateral transport of hydrothermal fluids (tens of km) does occur in young oceanic crust at mid-oceanic ridges (Davis and Becker, 1998). This circulation is driven mainly by the temperature gradient but also by hydrostatic pressure in some cases.

In areas adjacent to land, groundwater flow is likely to be driven by hydrostatic pressure and faults are likely to provide the main conduits for this flow (e.g. Glasby, 1998). Other factors may also influence groundwater flow in a coastal setting such as the hydraulic gradient, rock type and tidal pumping. In addition, magmatic fluid or vapour may be an important constituent at some sites located at convergent plate boundaries (Giggenbach 1992a,b; 1997). Figure 3 shows a simplified cross section of coastal hydrothermal circulation, where seawater, groundwater and magmatic water are mixed in various ratios under the influence of heat supplied by volcanic activity.

4.2. Hydrothermal fluid circulation and composition associated with multiple endmembers

The chemical composition and isotopic ratios of hydrothermal fluids are strongly dependent on the mixing of waters of different origins. Identification of the origin of the fluids in hydrothermally active coastal zones is important in elucidating the subsurface processes and fluid reservoir characteristics.

There are a number of possibilities for the secondary modification of venting fluids (Gamo, 1995). Phase separation is a common feature in shallow hydrothermal systems, not only because lower hydrostatic pressure facilitates boiling of the ascending hydrothermal fluid but also because the fluid is originally rich in gas components with higher contents of magmatic (mantle-derived) volatiles. Entrained seawater becomes enriched in Cl^- as a result of phase separation but depleted in Cl^- as a result of the introduction of meteoric water (groundwater).

Since coastal zones are usually characterized by high sedimentation rates due to fluvial inflow and high primary productivity, hydrothermal circulation may cause thermocatalytic decomposition of organic material and carbonates within the sediment, resulting in significantly higher values of CO_2, CH_4, NH_4^+ and alkalinity in the hydrothermal fluid (Gamo, 1995).

In shallow systems, the water pressure and temperature are much lower than those at mid-oceanic ridges so that black smokers and hydrothermal chimneys do not form and the deposition of metals takes place mainly beneath the seafloor.

4.3. Associated biological communities

Abundant nutrient supply from shallow hydrothermal vents significantly raises biological production in coastal euphotic zones. This enhanced production chiefly takes the form of bacterial and diatom mats, not megafauna such as *Calyptogena* and tube worms (except for the existence of tube worms in Kagoshima Bay as described above) which are common at deep-sea hydrothermal sites. These specific hydrothermal communities in the coastal zones are based on both photosynthetic and chemosynthetic primary production. Fauna in these communities seems to depend on adaptations to chemically-modified habitats, such as increased concentrations of sulphides and trace metals.

The biogeochemical processes in coastal areas where hydrothermal activity is taking place differ significantly from those areas devoid of such activity. The simultaneous occurrence of the two primary production systems (photosynthesis and chemosynthesis) probably makes the functional characteristics of the coastal hydrothermal fauna more complicated than those in deep-sea hydrothermal systems (e.g. Tarasov et al., 1999). For example, the rate of photosynthesis may be depressed due to the high concentration of H_2S and trace metals directly over the vents. Recent studies on the diversity of bacteria and archaea at a coastal vent site in the Aegean Sea have shown that physicochemically different zones around a vent site are inhabited by distinct bacterial populations (Dando et al., 2000).

5. Summary

Previous studies have given a clearer understanding of the factors governing hydrothermal activity in the coastal zone. Detailed sampling of the vent fluids for chemical analysis together with topographic mapping and long-term monitoring of the venting fluids has allowed us to make an assessment of the importance of the shallow vents to the biogeochemical flux in various regions of the Mediterranean Sea and the western Pacific Ocean. It is worth noting that the CO_2 and CH_4 fluxes from the Aegean Sea may be of global significance suggesting that other coastal hydrothermal areas may also contribute significant amounts of CO_2 and CH_4 as well. It would therefore be desirable to estimate the global flux of many of the chemical components involved in coastal hydrothermal activity by conducting detailed chemical analysis of venting fluids, hydrothermal plumes and sedimentary pore fluids. Microbial activity related to both photosynthesis and chemosynthesis should also be an important target for the elucidation of biodiversity in the coastal hydrothermal conditions.

6. Acknowledgements

This study was partly supported by a grant from the JNC Cooperative Research Scheme on the Nuclear Fuel Cycle, a grant from the Ministry of Education, Culture, Sports, Science and Technology of Japan (No. 13304036) and by a Special Coordination Fund of MEXT "Archaean Park" project.

REFERENCES

Anon., 2001a. Global Volcanism Program Volcanic Activity Reports, Smithsonian National Museum of Natural History (http://www.nmnh.si.edu/gvp/gvn/index.htm).

Anon., 2001b. Earth, Oceans and Life. Scientific Investigation of the Earth System Using Multiple Drilling Platforms and New Technologies. Integrated Ocean Drilling Program Initial Science Plan, 110 pp. (http://www.iodp.org).

Botz, R., Stüben, D., Winckler, G., Bayer, R., Scmitt, M., Faber, E., 1996. Hydrothermal gases from offshore Milos Island, Greece. Chem. Geol. 130, 161-173.

Craig, H., Horibe, Y., 1994. ^{3}He and methane in Sakurajima Caldera, Kagoshima Bay, Japan, Earth Planet. Sci. Lett. 123, 221-226.

Crossland, C.J., Kremer, H.H., 2002. Coastal Zones: Ecosystems under Pressure. Paper presented to the Global Conference on Oceans and Coasts at Rio+10, towards the 2002 World Summit on Sustainable Development, Johannesburg. IOC UNESCO, Paris, December 3-7 (in preparation).

Dando, P.R., Hughes, J.A., Leahy, Y., Taylor L.J., Zivanovic, S., 1995a. Earthquakes increase hydrothermal venting and nutrient inputs into the Aegean. Cont. Shelf Res. 15, 655-662.

Dando, P.R., Hughes, J.A., Leahy, Y., Niven, S.J., Taylor L.J., Smith, C., 1995b. Gas venting rates from submarine hydrothermal areas around the island of Milos, Hellenic Volcanic Arc. Cont. Shelf Res. 15, 913-929.

Dando, P.R., Stüben, D., Varnavas, S.P., 1999. Hydrothermalism in the Mediterranean Sea. Prog. Oceanogr. 44, 333-367.

Dando, P.R., Aliani, S., Arab, H., Bianchi, C.N., Brehmer, M., Cocito, S., Fowler, S.W., Gundersen, J., L., Hooper, E., Kölbl, R., Kuever, J., Linke, P., Makropoulos, K.C., Meloni, R., Miquel, J.-C., Morri, C., Müller, S., Robinson, C., Schlesner, H., Sievert, S., Stöhr, R., Stüben, D., Thomm, M., Varnavas S.P., Zeibis, W., 2000. Hydrothermal studies in the Aegean Sea. Phys. Chem. Earth (B) 25, 1-8.

Davis, E., Becker, K., 1998. Borehole observations record driving forces for hydrothermal circulation in young oceanic crust. EOS Trans. Am. Geophys. Union 79(31), 369-370.

de Ronde, C.E.J., 1995. Fluid chemistry and isotopic characteristics of seafloor hydrothermal systems and associated VMS Deposits: Potential for magmatic contributions. Mineral. Assoc. Canada Short Course Ser. 23, 479-509.

Elderfield, H., Schulz, A., 1996. Mid-ocean ridge hydrothermal fluxes and the chemical composition of the ocean. Ann. Rev. Earth Planet. Sci. 24, 191-224.

Gamo, T., 1995. Wide variation of chemical characteristics of submarine hydrothermal fluids due to the secondary modification processes after high temperature water-rock interaction: a review. In: Sakai, H., Nozaki, Y. (Eds.), Biogeochemical Processes and Ocean Flux in the Western Pacific, Terra Scientific Publishing Company (TERRAPUB), Tokyo, pp. 425-451.

Gamo, T., Okamura, K., Charlou, J-L., Urabe, T., Auzende, J-M., Ishibashi, J., Shitashima, K., Chiba, H., Shipboard Scientific Party of the ManusFlux Cruise, 1997. Acidic and sulfate-rich hydrothermal fluids from the Manus back-arc basin, Papua New Guinea. Geology 25, 139-142.

Gerlach, T.M., 1989. Degassing of carbon dioxide from basaltic magma at spreading centres: II. Mid-oceanic ridge basalts. J. Volcanol. Geothermal. Res. 39, 221-232.

Giggenbach, W.F., 1992a. Magma degassing and mineral deposition in hydrothermal systems along convergent plate boundaries. Econ. Geol. 87, 1927-1944.

Giggenbach, W.F., 1992b. Isotopic shifts in waters from geothermal and volcanic systems along convergent plate boundaries and their origin. Earth Planet. Sci. Letts 113, 495-510.

Giggenbach, W.F., 1997. The origin and evolution of fluids in magmatic hydrothermal systems. In: Barnes, H.L. (Ed.), Geochemistry of Hydrothermal Ore Deposits Third Edition, John Wiley & Sons, Inc., N.Y., pp. 737-796.

Glasby, G.P., 1998. The relation between earthquakes, faulting and submarine hydrothermal mineralization. Mar. Georesourc. Geotechnol. 16, 145-175.

Hashimoto, J., Miura, T., Fujikura, K., Ossaka, J., 1993. Discovery of vestimentiferan tube-worms in the euphotic zone. Zool. Sci. 10, 1063-1067.

Heikoop, J.M., Tsujita, C.J., Risk, M.J., Tomascik, T., Mah, A.J., 1996. Modern iron ooids from a shallow-marine volcanic setting: Mahengetang, Indonesia. Geology 24, 759-762.

Herzig, P.M., Hannington, M.D., 2000. Input from the deep: hot vents and cold seeps. In: Schulz, H.D., Zabel, M. (Eds.), Marine Geochemistry, Springer-Verlag, Berlin, pp. 398-416.

Holligan, P.M., de Booijs, H., 1993. Land-sea interactions in the coastal zone: science plan. The International Geosphere-Biosphere Programme (IGBP) Report No. 25 (Stockholm).

Horibe, Y., Gamo, T., Baba, Y., 1980. Estimation of carbon dioxide flux from submarine volcano in Kagoshima Bay, Japan. In: Goldberg, E.D., Horibe, Y., Saruhashi, K. (Eds.), Isotope Marine Chemistry, Uchida Rokakuho Pub. Co. Ltd., Tokyo, pp. 327-337.

Kadko, D., Baross J., Alt, J., 1995. The magnitude and global implications of hydrothermal flux. In: Humphris, S.E., Zierenberg, R.A., Mullineaux, L.S., Thomson, R.E. (Eds.), Seafloor Hydrothermal Systems: Physical, Chemical, Biological, and Geological Interactions. Geophysical Monograph 91. American Geophysical Union, pp. 446-466.

Kagoshima Prefecture, 1978. Report on the environmental survey for the mercury pollution in Kagoshima Bay. 38 pp. (in Japanese).

Kamada, M., Sakamoto, H., Yonehara, N., Ohnishi, T., 1977. Anomalous concentrations of chemical components in sea water of Kagoshima Bay. In: Kamada, M. (Ed.), Submarine Volcanic Activities and Abnormal Marine Environments at Northern Sakurazima Area, pp. 64-80 (in Japanese, English abstract).

Kerrick, D.M., McKibben, M.A., Seward, T.M., Caldeira, K., 1995. Convective hydrothermal CO_2 emission from high heat flow regions. Chem. Geol. 121, 285-293.

Marty, B., Jambon, A., Sano, Y., 1989. Helium isotopes and CO_2 in volcanic gases in Japan. Chem. Geol. 76, 25-40.

Moore, G.F., Taira, A. Klaus, A. et al., 2001. Deformation and fluid processes in the Nankai Trough accretionary prism sites 1173-1178. Proc. ODP, Init. Repts., Vol. 190. (http://www.odp.tamu.edu).

Moore, T., Brewer, P., 2001. Ocean Sciences at the New Millenium. National Science Foundation, Washington,D.C. 160 pp. (http://www.joss.ucar.edu/joss_psg/publications/).

Nagao, K., Kita, I., Matsuda, J., Mitripoulos, P., 1991. Noble gas isotope geochemistry of volcanic gases from the Aegean Island Arc. Bull. Geol. Soc. Greece 25, 33-41.

Ossaka, J., Hirabayashi, J., Nogami, K., Kurosaki, M., Hashimoto, J., 1992. Variation of chemical composition of volcanic gases from the northern part of Kagoshima bay, Proc. JAMSTEC Symp. Deep Sea Res. 8, 75-80 (in Japanese with English abstract).

Pantin, H.M., Wright, I.C., 1994. Submarine hydrothermal activity within the Taupo Volcanic Zone, Bay of Plenty continental shelf, New Zealand. Cont. Shelf Res. 14, 1411-1438.

Pernetta, J.C., Milliman, J.D., 1995. Land-Ocean Interactions in the Coastal Zone --- Implementation Plan. IGBP Global Change Report No. 33. International Geosphere-Biosphere Programme, Stockholm, Sweden. 215 pp.

Pichler, T., Veizer, J., 1999. Precipitation of Fe(III) oxyhydroxide deposits from shallow-water hydrothermal fluids in Tutum Bay, Ambitle Island, Papua New Guinea. Chem. Geol. 162, 15-31.

Pichler, T., Veizer, J., Hall, G.E.M., 1999. The chemical composition of shallow-water hydrothermal fluids in Tutum Bay, Ambitle Island, Papua New Guinea and their effect on ambient seawater. Mar. Chem. 64, 229-252.

Sakai, H., Matsubaya, O., 1974. Isotope geochemistry of the thermal waters of Japan and its bearing on the Kuroko ore solutions. Econ. Geol. 69, 974-991.

Sakai, H., Matsubaya, O., 1977. Stable isotopic studies of Japanese geothermal systems. Geothermics 5, 97-124.

Sarano, F., Murphy, R.C., Houghton, B.F., Hedenquist, J.W., 1989. Preliminary observations of submarine geothermal activity in the vicinity of White Island Volcano, Taupo Volcanic Zone, New Zealand. Jl R. Soc. N.Z. 4, 449-459.

Sedwick, P., Stüben, D., 1996. Chemistry of shallow submarine warm springs in an arc-volcanic setting: Vulcano Island, Aeolian Archipelago, Italy. Mar. Chem. 53, 147-161.

Stüben, D., Sedwick, P. Colantoni, P., 1996. Geochemistry of submarine warm springs in the limestone cavern of Grotta Azzurra, Capo Palinuro, Italy: Evidence for mixing-zone dolomitisation. Chem. Geol. 131, 113-125.

Stüben, D., Glasby, G.P., 1999. Geochemistry of shallow submarine hydrothermal fluids from Paleohori Bay, Milos, Aegean Sea. Explor. Mining Geol. 8, 273-287.

Tarasov, V.G., Gebruk, A.V., Shulkin, V.M., Kamenev, G.M., Fadeev, V.I., Kosmynin, V.N., Malakhov, V.V., Starynin, D.A., Obzhirov, A.I., 1999. Effect of shallow-water hydrothermal venting on the biota of Matupi Harbour (Rabaul Caldera, New Britain Island, Papua New Guinea). Cont. Shelf Res. 19, 79-116.

Tarasov, V.G., Propp, M.V., Propp, L.N., Zhirmunsky, A.V., Namsaraev, B.B., Gorlenko, V.M., Starynin, D.A., 1990. Shallow-water gasohydrothermal vents of Ushishir Volcano and the ecosystem of Kraternaya Bight (The Kurile Islands). Mar. Ecol. 11, 1-23.

Thompson, G., 1983. Hydrothermal fluxes in the ocean. In: Riley, J.P., Chester, R. (Eds.), Chemical Oceanography, vol. 8. Academic Press, London, pp. 271-337.

Vidal, M.V., Vidal, F.V., Isaacs, J.D., Young, D.R., 1978. Coastal submarine hydrothermal activity off northern Baja California. J. Geophys. Res. 83, 1757-1774.

Vidal, M.V., Vidal, F.V., Isaacs, J.D., 1981. Coastal submarine hydrothermal activity off northern Baja California: 2. Evolutionary history and isotopic geochemistry. J. Geophys. Res. 86, 9451-9468.

Von Damm, K.L., 1990. Seafloor hydrothermal activity: black smoker chemistry and chimneys. Ann. Rev. Earth Planet. Sci. 18, 173-204.

Von Damm, K.L., 1995. Controls on the chemistry and temporal variability of seafloor hydrothermal fluids. In: Humphris, S.E., Zierenberg, R.A., Mullineaux, L.S., Thomson, R.E. (Eds.), Seafloor Hydrothermal Systems: Physical, Chemical, Biological, and Geological Interactions. Geophysical Monograph 91. American Geophysical Union, pp. 222-247.

Welhan, J.A., Craig, H., 1983. Methane, hydrogen and helium in hydrothermal fluids at 21°N on the East Pacific Rise. In: Rona, P.A., Boström, K., Laubier, L., Smith K.L. (Eds.), Hydrothermal processes at seafloor spreading centers. Plenum Press, N. Y., pp. 391-409.

Welhan, J.A., 1988. Origins of methane in hydrothermal systems. Chem. Geol. 71, 183-198.

Williams, S.N., Schaeffer, S.J., Calvache, M.L., Lopez D., 1992. Global carbon dioxide emission to the atmosphere by volcanoes. Geochim. Cosmochim. Acta 56, 1765-1770.

WRI, 2000. World Resources 2000-2001. People and Ecosystems. The Fraying Web of Life. World Resources Institute, Washington, D.C., 389 pp.

Yamanaka, T., Ishibashi, J., Kataoka, S., Kikawada, Y., Mizota, C., Hashimoto, J., 1999a. Geochemical studies of the hydrothermal system in Wakamiko Caldera, northern Kagoshima Bay. JAMSTEC J. Deep Sea Res. 15, 145-151 (in Japanese with English abstract).

Yamanaka, T., Mizota, C., Murae, T., Hashimoto, J., 1999b. A currently forming petroleum associated with hydrothermal mineralization in a submarine caldera, Kagoshima Bay, Japan. Geochem. J. 33, 355-367.

Yamanaka, T., Ishibashi, J., Hashimoto, J., 2000. Organic geochemistry of hydrothermal petroleum generated in the submarine Wakamiko caldera, southern Kyushu, Japan. Org. Geochem. 31, 1117-1132.

Yamanaka, T., Ishibashi, J., Nakano, A., Ueki, Y., Hashimoto, J., 2001. Geochemical study of seafloor hydrothermal system and chemosynthetic animal community around a submarine caldera, southern part of Kagoshima Bay, southern Kyushu, Japan. JAMSTEC J. Deep Sea Res. 15, 145-151 (in Japanese with English abstract).

Land and Marine Hydrogeology
Edited by M. Taniguchi, K. Wang and T. Gamo

High Permeability of Young Oceanic Crust Constrained by Thermal and Pressure Observations

Kelin Wang[a, b] and Earl E. Davis[a]

[a] Pacific Geoscience Centre, Geological Survey of Canada
9860 West Saanich Road, Sidney, B.C., Canada V8L 4B2
Tel: (250)363-6429; fax: (250)363-6565; e-mail: wang@pgc.nrcan.gc.ca

[b] School of Earth and Ocean Sciences, University of Victoria
P.O. Box 3055 STN CSC, Victoria, B.C., Canada V8W 3P6
Tel: (250) 721-6120; fax: (250) 721-6200

Water flow has long been recognized to take place in the oceanic crust, affecting its energy budget, mineral composition, and mechanical properties. Of fundamental importance is the permeability of the basaltic formation that determines the spatial dimension and the vigor of the flow. Along a transect on the Juan de Fuca Ridge flank perpendicular to the ridge, three types of thermal and pressure observations have been made to constrain the formation-scale permeability of the basaltic crust: (1) Seafloor heat flow and borehole temperature measurements reveal a nearly isothermal sediment-basement interface despite locally large basement topography and non-uniform sediment thickness. The thermal homogenization is unequivocally the result of vigorous buoyancy-driven convection within the igneous formation. (2) Formation fluid pressure varies in response to periodic seafloor loading, such as tidal loading. Observed pressure amplitudes and phases reveal a lateral pressure diffusion length scale of many kilometers in the igneous crust, indicating a very high permeability. (3) Continuous pressure records have captured the formation fluid pressure change in response to a local earthquake swarm, inferred to be an extension event along a segment of the Juan de Fuca Ridge. Subsequent pressure variations in different boreholes again indicate a diffusion length scale of many kilometers. Theoretical modeling of all three types of observations consistently yields formation permeabilities as high as 10^{-10} - 10^{-9} m^2, several orders of magnitude higher than values determined by direct measurements made in most deep sea boreholes. These observations and inferences further point to the scale-dependent nature of the permeability. The formation permeability is probably dominated by well connected fractures. Future challenges include determining the depth variation of the formation permeability and the depth extent of water circulation.

1. INTRODUCTION

Mafic rocks that constitute the igneous oceanic crust are created at mid-ocean ridge seafloor spreading centers as a byproduct of partial melt from upwelling mantle. Through various parts of the spreading process, permeability can be generated in both the intrusive gabbros and extrusive basalts. Extensional faulting creates interconnected fractures throughout the crust, and rapid quenching, degassing, faulting, and lava flow create breccia, talus, and other forms of rubble in the uppermost few hundred meters of the crustal section. The resultant permeability facilitates ventilated hydrothermal convection at the ridge crest and ridge flanks, and the cold sea water brought into the rock formation by this circulation further promotes fracture development by thermal contraction. Ventilated circulation is eventually prohibited by the deposition of sediments, but hydrothermal convection persists within the sediment-buried igneous rocks until permeability is reduced by void-filling mineralization. The permeability of old ocean crust can be rejuvenated by tectonic activities such as normal faulting and flexure at subduction zones.

Besides the obvious value of understanding economic mineral deposition, the permeability of oceanic crust is of interest for its role in plate tectonics. The oceanic crust should be "dry" when formed, in the sense that the pressure and temperature conditions of the ridge environment do not allow hydrous minerals to be stable except in the very shallow, chilled part. By the time the crustal rocks reach a subduction zone, they are "wet". Significant amounts of water are contained in hydrous alteration minerals produced where water has gained access to various levels in the crust. As the plate descends, increased pressure and temperature cause the basalt and gabbro to transform into a denser rock, eclogite. Dehydration of the altered crustal rocks in these metamorphic processes gives rise to volcanic arcs and probably earthquakes within the subducted crust (Kirby et al., 1996; Peacock and Wang, 1999). Dehydration of serpentine, a product of hydration of the upper mantle material, has also been proposed to be responsible for earthquakes that occur within the subducted oceanic mantle (Peacock, 2001). If this is true, some circulation must extend to sub-crustal depths. How much water gets into the oceanic crust and mantle and to what depth depend on the permeability structure.

In this article, we summarize our work in the past several years at the Juan de Fuca Ridge and its eastern flank and provide examples of how estimates of permeability of young oceanic crust have been derived. The Juan de Fuca Ridge is the divergent boundary between the Pacific and Juan de Fuca plates (Figure 1a), where new crust is created at an intermediate spreading rate of 56 mm/yr. The eastern flank eventually subducts beneath the North America plate at the Cascadia subduction zone, causing volcanoes and earthquakes. The western flank must travel a much greater distance before being eventually subducted at the Alaska/Aleutian subduction zone.

Most of the eastern flank is covered by low-permeability (~ 10^{-16} m^2) sediments (Snelgrove and Forster, 1996; Giambalvo et al., 2000), and thus the extrusive upper igneous crust forms a laterally extensive confined sub-seafloor aquifer (Figure 1b). The simple geological structure that involves few rock types and a simple hydrostatic seafloor boundary condition are ideal for the hydrologic studies described here.

Along a transect perpendicular to the ridge (Figure 1b), the sediment-basement boundary has been clearly mapped by reflection seismic surveys, the sub-seafloor thermal regime has been defined by detailed heat-flow measurements, and crustal temperatures and pressures have been determined in several Ocean Drilling Program (ODP) boreholes. Seafloor heat flow is measured with a probe driven by its own weight typically 3 - 4 m into the sediment. Temperatures and *in situ* thermal conductivities are measured at typically 30 cm intervals along the probe's length. The product of the thermal gradient and conductivity gives the local heat flux. The heat flux values thus determined along one profile are shown in Figure 2.

Boreholes have been drilled at a total of ten locations along the eastern flank transect. Four holes were cased through the sediment layer, deepened into the igneous basement, and sealed with

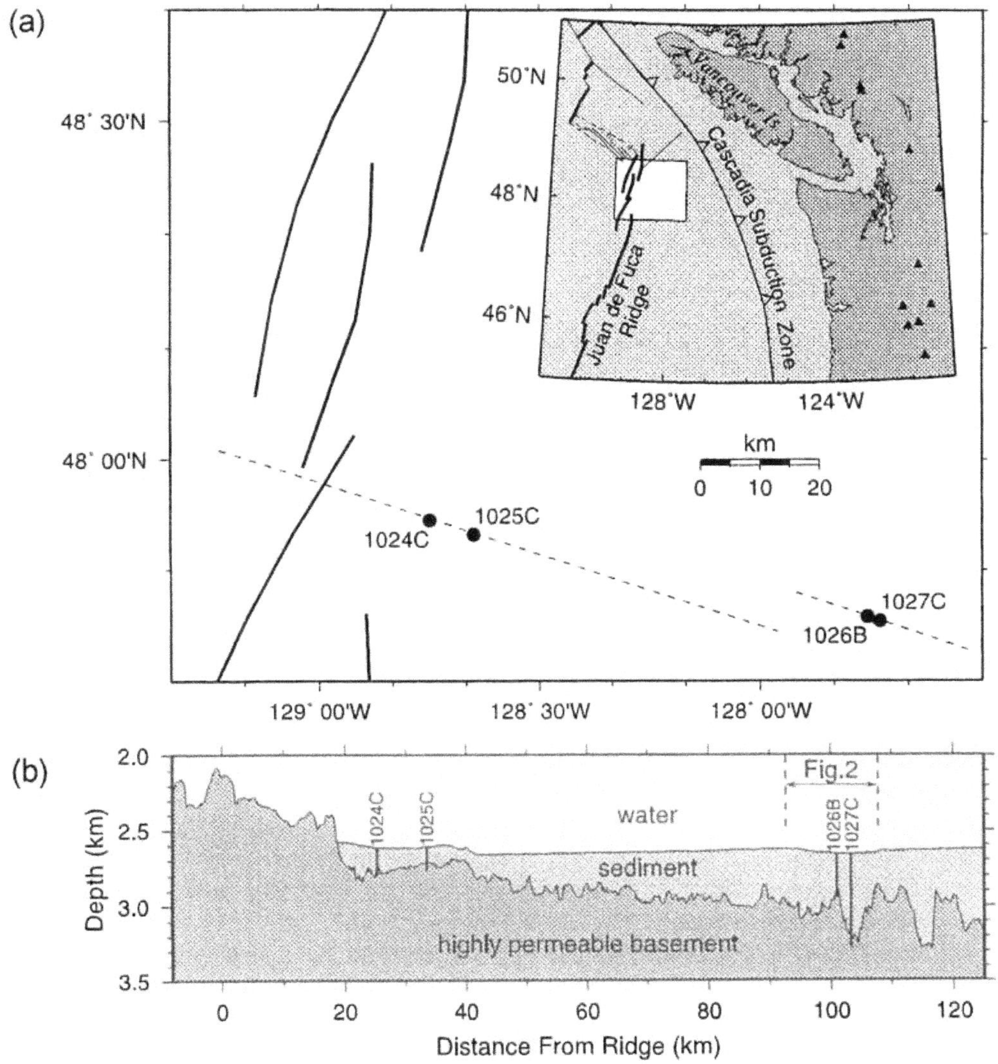

Figure 1. (a) Location of the geophysical transect (dashed line) discussed in this article and tectonic setting (inset). Solid circles are ODP boreholes. Triangles in the inset are active volcanoes. (b) Cross section along the transect showing seafloor and basement topography and locations of boreholes and heat flow measurements.

CORK (Circulation Obviation Retrofit Kit) instrumentation (Davis et al., 1992). CORKs prevent water from flowing into or out of the formation via the cased holes and thus allow a stable environment for monitoring temperature and fluid pressure and for sampling formation fluid. Pressure sensors are located at the seafloor and in the borehole just below the CORK seals. The high permeability of the upper crust allows efficient hydraulic communication between the formation and the pressure sensor in the cased borehole and permits any pressure change in the

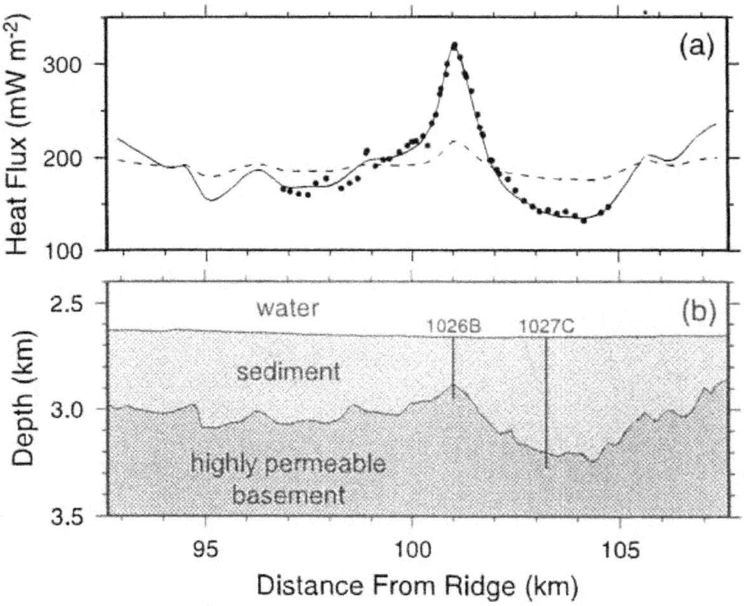

Figure 2. (a) Heat flow measurements (dots) along the profile shown in Figure 1b. Model predicted values (curves) are discussed in Section 2. (b) Cross section along the same profile.

permeable formation to be detected instantaneously; this point will be further addressed in Section 3. Pressure signals include contributions from several processes. (1) After the hole is sealed, the pressure gradually changes from the open hole value to the formation value. Whether the formation value is above or below hydrostatic depends on the hydrological environment of the borehole site. (2) As the oceanic crust is loaded by ocean tide, earth tide, or other oceanographic and barometric pressure variations (as recorded by the seafloor sensor), the formation pressure changes accordingly, with a reduced amplitude and a phase difference from the loading function. The most dominant components of this type of loading are the semi-diurnal and diurnal ocean tides driven by lunar and solar forcing. (3) Tectonic events such as seismic or aseismic faulting induce strain pulses in the rock formation. These in turn cause sudden changes in the interstitial fluid pressure which subsequently return to the unperturbed formation value. Examples of seafloor and formation pressure time series that include both the tidal signal and a tectonic signal are shown in Figure 3.

In the next three sections, we demonstrate how heat flow measurements, borehole temperature measurements, tidal pressure signals, and tectonic pressure signals can all be used to constrain the permeability of the uppermost igneous crust. Details of the measurements are given in various references we will subsequently quote, and in this article we focus on the theoretical inferences made from these measurements. Before we proceed, it is important to re-emphasize the scale dependence of permeability. It has long been noticed that permeability values generally increase with the scale of observations. Reviews have been given by Clauser (1992) for continental igneous rocks and by Fisher (1998) for the basaltic oceanic crust. If we take a small unfractured sample of basalt, there is virtually no permeability. A single borehole may encounter some small fractures, and the permeability inferred from instantaneous (slug) or continuous (injection) pumping tests

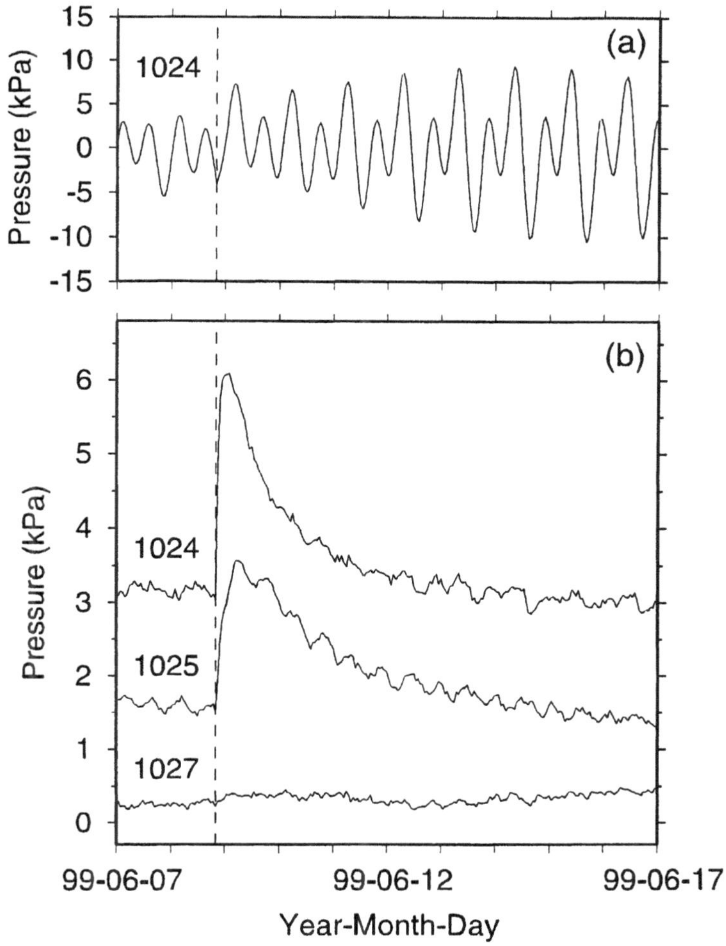

Figure 3. Examples of formation fluid pressure records. (a) Raw data from ODP hole 1024C (see Figure 1 for location) during June 7-17, 1999. The pressure signal is dominated by lunar and solar ocean tides. Implications of the tidal signals are discussed in Section 3. (b) Pressure records from holes 1024C, 1025C, and 1027C for the same time period after the tidal signals are removed. They show the pressure disturbance caused by a tectonic strain event of June 8, 1999. The tectonically induced pressure variation is discussed in Section 4. Modified from Davis et al. (2001).

may be representative of a region comparable to the size of the borehole. Permeability values determined this way are usually much larger than laboratory measurements made on small samples but are limited by the duration of the transient tests. The thermal and multi-borehole pressure measurements we cite in this article sample the crust over tens of kilometers, and the permeability we deal with is the "formation permeability". The formation permeability is probably dominated by large and well connected fractures. If there are large fractures, one may question the validity of regarding the formation as a porous medium at a scale of a few tens of meters, but at the full scale of the formation, a uniform porous medium is a reasonable approximation. Even in the limiting scenario in which the Darcy's law for fluid flow in porous media becomes very inaccurate, formation permeability is still a useful concept that allows the effective average hydrologic characteristics of a very permeable geologic formation to be quantified, in much the same way the "effective elastic thickness" is used to quantify the flexural strength of a lithospheric plate. The fluid flux, or the average rate of flow through a unit area oriented in a given direction in the porous medium, is called the Darcy velocity. An alternative term, the "specific discharge", is less preferred because it does not emphasize that the quantity has a direction (a vector).

2. HYDROTHERMAL CONVECTION

By extrapolating the heat flow measurements made in the Juan de Fuca Ridge flank to the sediment-basement interface, it is evident that the interface temperature is nearly constant over distances of several kilometers regardless of its topography and depth of sediment burial (Figure 2). This phenomenon in the Juan de Fuca Ridge flank was noticed by Davis et al. (1989), and they inferred that the interface was probably kept isothermal by vigorous water circulation in the igneous basement. Without water flow, basement topography combined with the thermal conductivity contrast between the sediment and the basement can only give rise to small heat flow variations at the seafloor, as shown by the dashed line in Figure 2. Temperature measurements in ODP holes 1026B and 1027C (Figure 1b) confirmed that, despite the 370 meter basement topographic relief and the 2.2 km separation of the sites, the interface temperatures at these two sites differed by less than 2 K (Davis and Becker, manuscript submitted to Earth Planet. Res. Lett., 2002). Hydrothermal convection appears to be the only explanation for the isothermality at the interface.

The vigor of buoyancy-driven thermal convection is characterized by the Rayleigh number Ra, a measure of thermal buoyancy vs. viscous resistance. For a layer having differential temperature dT across thickness h, the Rayleigh number is defined as

$$Ra = \frac{\kappa \rho^2 c \phi g h dT}{k\mu} = \frac{\kappa \rho^2 c \phi g h^2 \Gamma}{k\mu} \tag{1}$$

where κ is permeability, g is gravitational acceleration, k is the thermal conductivity of the fluid-saturated porous medium, ρ, c, ϕ, and μ are the density, specific heat, thermal expansion coefficient, and dynamic viscosity, respectively, of the fluid, and $\Gamma = dT/h$ is the average thermal gradient across the layer. None of these parameters is expected to be uniform in the real oceanic crust, but there are few constraints on the nature of variations and we usually use regional average values. Among all the parameters, permeability has the widest range of variation in nature and is least well known. Estimates for the permeability of the basaltic crust vary by at least fourteen orders of magnitude (Fisher, 1998). Therefore the vigor of thermal convection is most strongly affected by the permeability of the hosting formation. Other parameters do vary, but to a much lesser degree. At a higher temperature, water has greater thermal expansion and is less viscous, and hence Ra is higher. A thicker layer or a higher geothermal gradient also promotes convection.

For a flat layer with uniform physical properties, the system becomes unstable when Ra is greater than a critical value Ra_c of about 40, and thermal convection takes place. Subcritical convection does occur if there is lateral thermal heterogeneity, as long as Γ is greater than adiabatic. In fact, it must occur everywhere in the oceanic crust since there is always some seafloor and basement topography that causes thermal heterogeneity. Such convection is very slow and has very little thermal effect, but it may have important geochemical consequences if the flow persists over geological time scales. In a two-dimensional (2-D) situation (as is simulated in an experimental Hele-Shaw cell), convection will not be steady state if Ra is greater than about 10 Ra_c. The change of convection pattern with time will be either periodic or, at even higher permeabilities, chaotic (Wang et al., 1997).

At what κ and hence Ra values does convection make the sediment-basement interface at our observation site isothermal? The degree of isothermality of the sediment-basement interface is best described by the maximum temperature difference ΔT between any two points along the interface. We first use a 2-D thermal convection model, with an idealized sinusoidal sediment-basement interface, to investigate how permeability affects the vigor of convection and ΔT and the condition under which ΔT is reduced to less than 2 K. Figure 4a illustrates the model geometry, and the figure caption gives the values of material properties used. The model is designed not to reproduce a specific ridge flank setting but to explore the physical process. The wavelength of the sinusoidal sediment-basement interface is similar to the dominant topography of the study area (Figure 1b). A thickness of 600 m is used for the convection layer (layer 2 in Figure 4a), based on the presence of a seismic reflector 600 m below the interface at the Juan de Fuca Ridge flank (Rohr, 1994). Such a reflector is widely observed in the oceanic crust at depths of a few to several hundred meters. It is interpreted as the boundary between the extrusive and intrusive igneous rocks, and therefore as a boundary of major permeability contrast (Houtz and Ewing, 1976; Wilkens et al., 1991). Below this depth, there still is permeability and water circulation, but very vigorous convection is unlikely to occur. All water properties as functions of pressure and temperature are derived from the steam tables of Keenan et al. (1978). The calculated ΔT as a function of the middle-layer permeability is shown in Figure 4b. More details are given by Davis et al. (1997).

With low permeability values ($< 4 \times 10^{-15}$ m^2), Ra is subcritical, and the thermal regime is nearly purely conductive. Topographically induced subcritical convection has very little thermal effect. The basement trough is more deeply buried by the sediment and is warmer than the basement peak, which gives a ΔT of about 40 K. When the permeabilities are greater than a critical value of about 4×10^{-15} m^2, free convection takes place and convection cells form in the permeable layer. The sediment-basement interface is warmer where the flow is upward and colder where flow is downward, and the resultant ΔT is larger than that of the purely conductive regime. The ΔT increase with permeability is not smooth because the number of convection cells may change with a slight increase in permeability, resulting in an abrupt change in ΔT. At even higher permeabilities ($> 10^{-13}$ m^2), ΔT decreases with increasing permeability because the more vigorous convection tends to redistribute heat more effectively. The decreasing trend indicates that at very high permeabilities the water will be so well mixed that the basement will be nearly isothermal (or more strictly, nearly adiabatic, meaning nearly no conductive heat exchange within the layer). At $\kappa >$ 10^{-12} m^2, when Ra is about 10 times the critical value, steady state solutions generally no longer exist. However, by extrapolating the smooth decreasing trend of ΔT and considering the results of two runs that did yield stable solutions at high permeabilities, one can predict that ΔT will not be less than 2 K until κ is 10^{-10} - 10^{-9} m^2. This gives a Rayleigh number of at least 4000, which means the convection must be chaotic. The peak to trough depth difference of the sediment-basement interface in the simple model is 200 m, but the maximum depth difference for the dominant topography of the study area is 370 m. Since a greater topographic relief requires a more vigorous convection in order to be kept isothermal to the same degree, the model provides a

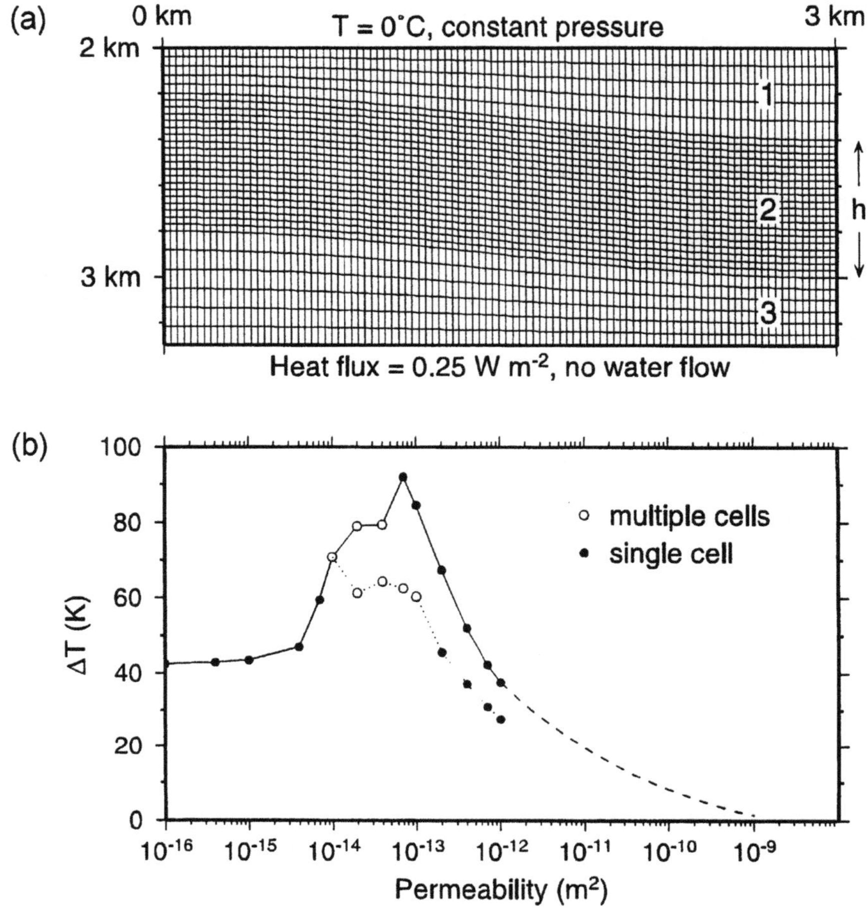

Figure 4. (a) The grid and boundary conditions of the finite element model used to simulate hydrothermal convection. The 3-km half-wavelength sinusoidal basement relief approximates the dominant basement topography along the heat flow profile (Figure 2b). The two vertical boundaries are boundaries of symmetry (i.e., no heat or fluid flow across them). Thermal conductivity is 1 W m^{-1} K^{-1} for layer 1 (representing sediment) and 2 W m^{-1} K^{-1} for layers 2 and 3; permeability is 10^{-16} m^2 for layer 1 and 10^{-17} m^2 for layer 3. (b) Maximum temperature variation (ΔT) along the model sediment-basement interface calculated using different permeabilities for the middle layer. Depending on the permeability, the pattern of convection in the middle layer alternates between a single convection cell and multiple cells. Points connected by the solid line are results obtained using a "normal" initial condition that would facilitate upward flow under the basement peak. Points connected by the dotted line are results obtained using a initial condition that would facilitate downward flow under the basement peak. ΔT increases from very low to moderately high permeabilities, but decreases at higher permeabilities. Steady state solutions do not exist when the permeability is higher than 10^{-12} m^2. The dashed line shows the visually extrapolated trend of the ΔT decrease in the permeability region of time-varying convection. Both panels are modified from Davis et al. (1997).

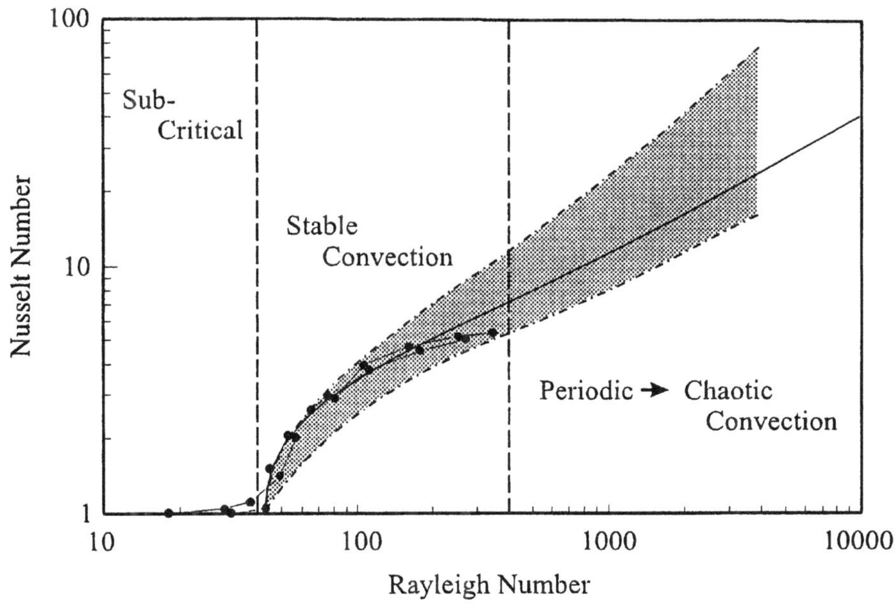

Figure 5. Nusselt number as a function of Rayleigh number for a porous layer heated from below. The range of laboratory and numerical modeling results is shown by the stippled area, and a relationship given by Lister (1990) (equation (3) in text) is shown by the solid line. Solid circles represent hydrothermal convection modeling results reported by Davis et al. (1997). From Davis et al. (1997).

conservative (lower limit) estimate for the formation permeability.

For a highly unsteady system, it is not very helpful to try to model the exact time-dependent convection patterns. Instead, we may understand the overall behavior of the convection system best by examining its average thermal budget. The thermal budget for the convecting layer that we used to define *Ra* is described by the Nusselt number *Nu*, defined as the ratio of the total heat transferred (Q) to the heat that would be transferred by conduction alone in the absence of convection:

$$Nu = \frac{Q}{k\Gamma} \tag{2}$$

Averaged over a large area, the total heat Q that comes from the earth's interior and passes through the layer to reach the seafloor does not change no matter how the system convects, but more vigorous convection causes a larger fraction of this heat to be transferred through the layer by convection. Therefore, *Nu* increases with *Ra*. In the extreme case in which the thermal gradient is adiabatic in the layer, all heat is transferred by convection. The relation between *Nu* and *Ra* is a subject of fundamental importance and great fascination. Figure 5 shows the range of experimental and numerical modeling results for the *Nu* - *Ra* relation, together with a function given by Lister (1990):

$$Nu = 0.16(Ra + 700)^{0.25}(Ra - 4\pi^2)^{0.4} \tag{3}$$

From (2), the total heat transfer through the convecting layer is $Q = Nu(k\Gamma)$. This is to say, in the presence of very vigorous convection, the thermal effect can be equivalently described using a purely conductive regime with a thermal conductivity that is Nu times the actual conductivity (Figure 6). A finite element conduction model is readily constructed using the real basement topography of Figure 1b and 2 (Davis et al., 1997). The model-predicted maximum temperature difference ΔT along the sediment-basement interface as a function of Nu is shown in Figure 7 for a few assumed thicknesses of the convection layer. To have an isothermal (to within 2 K) sediment-basement interface for a thickness of 600 m, we need to use a thermal conductivity 100 times the actual value of 2 W $m^{-1}K^{-1}$, that is, $Nu = 100$. Using the Lister Nu - Ra relation [equation (3)], we can infer that Ra has to be greater than 10000, which in turn gives a formation permeability of nearly 10^{-9} m^2. The surface heat flow calculated from such a simple model agrees with the observed values almost exactly (Figure 2).

3. TIDALLY INDUCED PRESSURE VARIATION AND FLUID FLOW

To understand how tidally induced fluid pressure variations can be used to constrain the formation permeability, we first briefly describe how water flows in a poroelastic medium under time-varying loading. The sediment or, at a formation scale, the permeable igneous basement can be considered to consist of a solid matrix hosting a fluid in its pore spaces (Figure 8a). If there is a change in external forcing such as due to ocean tides, both the matrix and pore fluid deform elastically. Because the matrix and the fluid have different elastic properties, they respond to the loading differently. The less compressible material, usually the matrix, develops more internal stress to take up the load. The more compressible material, usually the fluid, takes up less load. How the total load is instantaneously divided between the matrix and fluid depends also on porosity (n). A higher porosity enhances the effect of fluid compressibility. The fraction of the total incremental pressure change (i.e., the average of the three principal incremental stresses) instantaneously (elastically) taken up by the fluid is called the loading efficiency (γ). If there is no horizontal deformation, as is usually assumed for tidal loading, the loading efficiency becomes the one-dimensional "tidal" loading efficiency (γ'). For seawater, $\gamma = \gamma' = 1$, since there is no matrix frame to share any of the load. For a completely rigid matrix, $\gamma = \gamma' = 0$, since the matrix would take up all the load by itself.

The loading efficiency as an elastic property of the poroelastic medium is defined in Table 1. Other equivalent but lengthier expressions for them have been given by van der Kamp and Gale (1983). One can best understand the physical meaning of the quantity by considering a limiting situation in which the solid constituents are assumed incompressible (compressibility $\beta_s = 0$). In this case

$$\gamma = \frac{1}{1 + n(\beta_f / \beta)} \tag{4}$$

where β and β_f are the compressibilities of the matrix and fluid, respectively. A more compressible

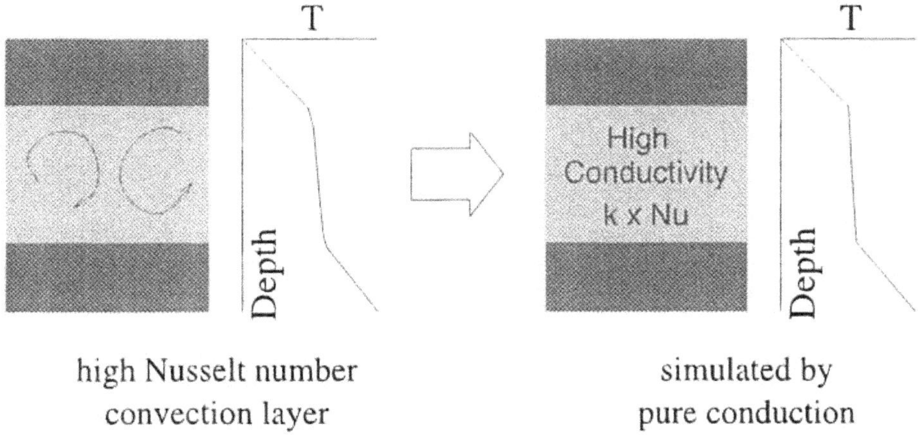

Figure 6. Schematic illustration to show how the thermal effect of very vigorous convection can be approximated using a model of pure conduction. In the real situation, the actual thermal conductivity of the layer is k, the Nusselt number characterizing convective transfer is Nu. In the approximation, heat transfer including convection is simulated using a conductivity of $k \times Nu$ in a conduction model.

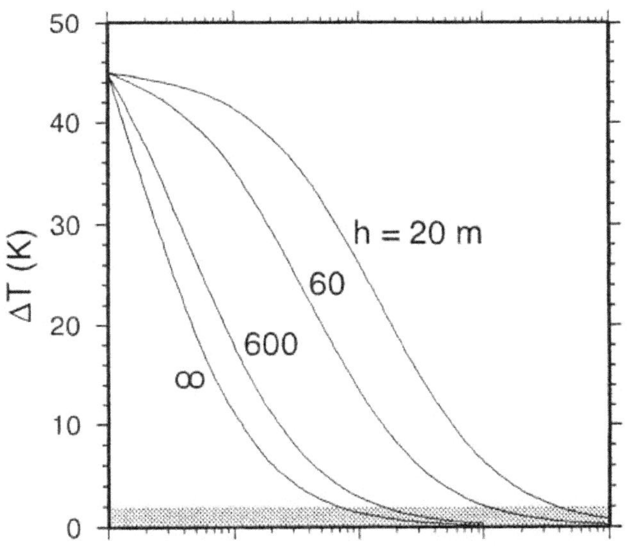

Figure 7. Maximum variation in temperature along the sediment-basement interface as a function of Nu for a purely conductive finite element model for the profile shown in Figure 2b, using the real basement topography. A high Nu simulates the effect of vigorous hydrothermal circulation (Figure 6). Predicted surface heat flux for a 600-m layer with $Nu = 100$ is shown in Figure 2a (solid line).

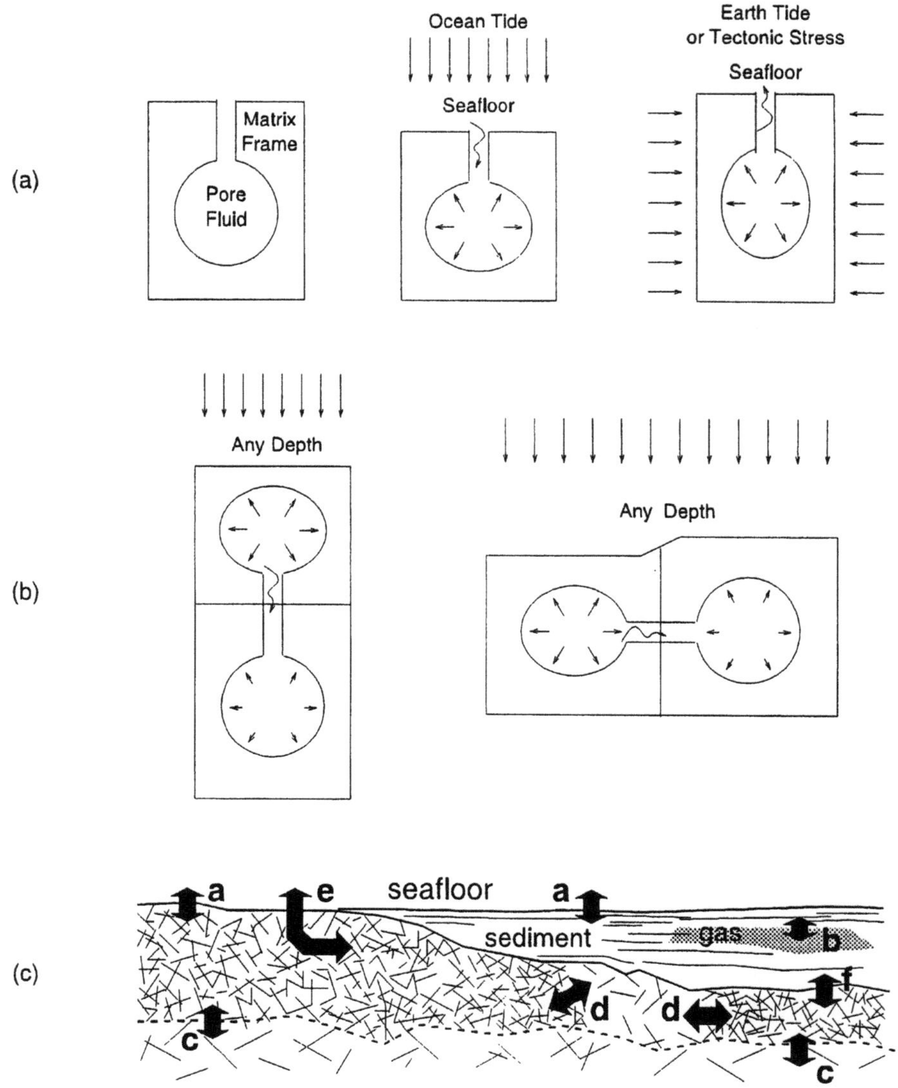

Figure 8. (a) A subsea formation can be considered to be a poroelastic medium consisting of a solid matrix frame and a pore fluid. When the medium is compressed, the fluid and frame "share" the load, and the fraction taken up by the fluid is the loading efficiency. The loading efficiency is 1 (i.e., 100%) above seafloor where no solid matrix shares the load with water but is less than 1 below the seafloor. The differential pressure across the seafloor induces Darcy flow (the wiggly arrow), adding a diffusive component to the pore fluid pressure response. (b) Wherever there is a loading efficiency contrast, differential elastic fluid pressures tend to induce fluid flow. (c) Formation boundaries, fracture density and porosity variations, and changes in fluid elastic properties all promote tidally induced fluid flow in subsea formations. Letter-labeled arrows represent examples of tidal flow: **a** and **e** – flow across the seafloor; **b** – flow at boundaries of gas-rich sediments, **c** and **d** – flow between regions of different frame compressibilities.

matrix leads to a larger loading efficiency because the fluid has to take up a greater share of the incremental load. While it appears that $\gamma < 1$ when $n = 1$ (no solid matrix) according to (4), in fact, $\gamma = 1$ in this case because β would approach infinity if porosity approaches unity. The expression for the corresponding (1-D) tidal loading efficiency is

$$\gamma' = \frac{1}{1 + n(\beta_f / \beta')} \tag{5}$$

where β', defined in Table 1, is the 1-D, or (horizontally) confined, compressibility of the matrix.

Different elastic responses in different parts of the geological formation give rise to non-hydrostatic fluid pressure gradients. For example, at the seafloor, γ or γ' abruptly changes from 1 in the seawater to a smaller value in the formation, and the resultant pressure gradient tends to drive the fluid to flow across the boundary (Figure 8a). Differential pressure is also created within the poroelastic medium wherever there is a contrast in loading efficiency (Figure 8b). Whether fluid flow actually takes place depends on the type of loading. If the loading is periodic and the frequency is high, as is usually the case when a seismic wave passes through the medium, the elastic pressure gradient is reversed before the fluid has time to flow. As a result, the loading

Table 1
Definition of elastic properties.

Symbol*				Name	Other Names	Definition†
(1)	(2)	(3)	(4)			
ν	ν	ν	ν	frame Poisson's ratio	drained Poisson's ratio	
β	$1/K$	$1/K_f$	$1/K$	frame compressibility	drained compressibility	
β'	$1/K'$	$1/K_f'$	$1/K'$	1-D frame compressibility	confined frame compressibility	$\beta(1+\nu)/[3(1-\nu)]$
β_f	$1/K_f$	$1/K_w$	$1/K_f$	fluid compressibility		
β_s	$1/K_s$	$1/K_s$	$1/K_s$	frame constituent compressibility	solid compressibility	
α	α	α	α	Biot-Willis parameter	effective-stress coefficient	$1 - \beta_s/\beta$
α_f	—	—	—			$1 - \beta_s/\beta_f$
λ	λ	λ	λ			$2\alpha(1-2\nu)/[3(1-\nu)]$
ζ	$S'/\rho g$	—	—	storage compressibility		$\alpha\beta + n\alpha_f\beta_f$
ζ'	$S_s/\rho g$	S	S	1-D storage compressibility		$\zeta - \alpha\beta\lambda$
γ	β	β	—	loading efficiency	Skempton's B coefficient	$\alpha\beta/\zeta$
γ'	γ	γ	γ	1-D loading efficiency	tidal loading efficiency	$\alpha\beta'/\zeta'$

* Different symbols have been used for the same parameter in various publications: (1) present article, (2) van der Kamp and Gale (1983), (3) Wang and Davis (1996), and (4) Wang et al. (1998).

† In these expressions, n is porosity.

effectively causes no fluid flow. Under such "undrained" conditions, the fluid pressure is simply the total pressure times the loading efficiency. However, when the loading frequency is low as in the case of tidal loading, the fluid may be able to flow in the direction of the local pressure gradient and hence modify the pressure field. The flow is governed by Darcy's law, and the fluid pressure change associated with this flow is called the diffusive component of the pressure variation. As with all diffusion processes, the "speed" of the diffusion is quantified by the diffusivity. Hydraulic diffusivity η depends on permeability, fluid viscosity, the elastic properties of the constituents of poroelastic medium, and whether matrix deformation is 3-D or 1-D (e.g., horizontally confined). For 3-D deformation

$$\eta = \frac{\kappa}{\mu\zeta} \tag{6}$$

where μ is the dynamic viscosity of the fluid, and ζ is the 3-D storage compressibility of the medium. In a 1-D case, ζ in (6) is replaced with the 1-D storage compressibility ζ'. For $\beta_s = 0$, $\zeta = \beta + n\beta_f$, and $\zeta' = \beta' + n\beta_f$. The exact expressions for ζ and ζ' without assuming $\beta_s = 0$ are listed in Table 1. ζ' is the commonly used specific storage coefficient divided by ρg, and ζ is its 3-D version. We deliberately avoid involving ρg to emphasize that hydraulic diffusivity has nothing to do with density and gravity.

If there is a sudden pressure increase p_o at a boundary, the time (τ) it takes for the diffusive component of the pressure variation at a distance L from the boundary to increase from zero to 48% [i.e., erfc(0.5)] of p_o is called the diffusion time constant for length scale L:

$$\tau = \frac{L^2}{\eta} \tag{7}$$

If the pressure change at the boundary is a sinusoidal function of time with period F, the diffusive component of pressure signal propagates into the medium as a diffusion wave, with the amplitude decreasing exponentially with distance. Half a wavelength from the boundary, the amplitude of the diffusion wave attenuates by a factor of $e^{-\pi}$. This distance is called the penetration length of the pressure wave and is defined as

$$q = \sqrt{\pi\eta F} \tag{8}$$

The penetration length of a diffusion wave decreases with decreasing diffusivity and wave period. For most seismic wave frequencies, q is practically zero, although in very permeable materials such as the upper igneous crust and coarse-grain sandy sediments, q can reach tens of meters at a period of one-second. Any loading-efficiency contrast generates a pressure gradient if there is a temporal stress change. Hence, diffusion waves can be generated at the seafloor or any boundary in the formation across which there is a loading efficiency contrast. The oceanic crust is heterogeneous at the local scale, and hence tidally induced fluid flow is widely present in subsea formations (Figure 8c). Details of pressure wave generation, propagation, and reflection around boundaries of loading efficiency contrast are considered by Wang and Davis (1996) and Wang et al. (1998). Energy dissipation by the tidally induced diffusive flow near these boundaries is discussed by Wang et al. (1999).

Given a sinusoidal tidal loading with amplitude σ_b at the surface of a poroelastic half-space, the fluid pressure we would observe within the medium is a combination of the elastic response and the diffusive response, as is shown in Figure 9. The elastic response is instantaneous, that is, always in phase with the loading function, and, assuming no horizontal deformation, its amplitude is $\gamma'\sigma_b$. The diffusive component has a phase lag because it takes time for the diffusion wave to propagate from the loading boundary. For a uniform half-space, there is no phase lag at depths that are multiples of half-wavelength q, as is expected of wave propagation. The amplitude of the diffusive component is $(1 - \gamma')\sigma_b e^{-\pi z/q}$. So in a uniform half-space, the formation response is practically elastic at depths much greater than q. The amplitude and phase of the total pressure signal can be derived from those of the elastic and diffusive components (Wang and Davis, 1996). In Figure 10, we show the amplitude and phase for the total fluid pressure in a uniform half-space under 12-hour period tidal loading, assuming a permeability of 1.7×10^{-10} m^2. All other medium properties assumed are given in Table 2. The position of the zero cross-over of the phase curve gives the q value. The permeability of seafloor sediment is typically around 10^{-16} m^2, so its q for the 12-hour tide is only a few meters, three orders of magnitude less than what is shown in Figure 10.

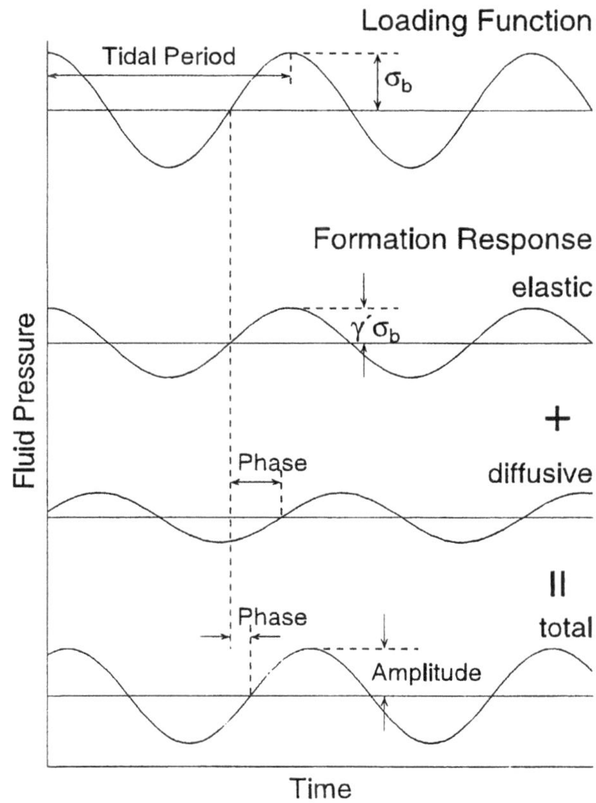

Figure 9. Schematic illustration of pressure variations due to tidal loading.

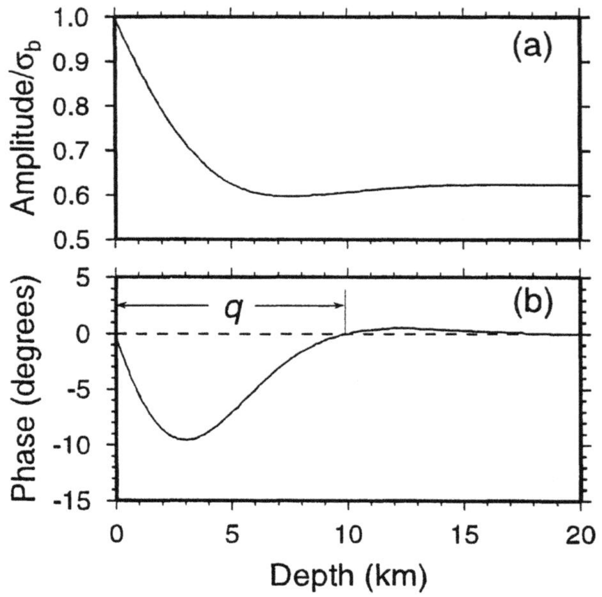

Figure 10. Pressure amplitude and phase as functions of depth for a homogeneous poroelastic half-space under 12-hr tidal loading at the surface. The permeability is assumed to be 1.7×10^{-10} m^2, and other parameters are given in Table 2.

The transect where we have obtained our formation tidal pressure observations does not resemble a half-space, but the half-space solution can be applied in the following manner. At the western end of the borehole transect (Figure 1b), the uncovered part of the permeable igneous basement allows efficient hydraulic communication between the formation and seawater. Water flows into and out of the basement outcrop, and the diffusive pressure wave propagates horizontally into the sediment-covered part of the basement to the east, as illustrated by the pattern marked "e" in Figure 8c. The simplified geometry is shown in Figure 11a. Note that where basement is covered by the 100 to a few hundred meters of sediment, a diffusive pressure wave cannot propagate from the seafloor locally downward to the basement because of the small

Table 2
Parameter values used in the models of Sections 3 and 4.

Symbol	Name	Value
β	matrix frame compressibility	3.33×10^{-10} Pa^{-1}
β_f	fluid compressibility	4.17×10^{-10} Pa^{-1}
β_s	matrix constituents compressibility	0.2×10^{-10} Pa^{-1}
n	porosity	0.2
μ	dynamic viscosity of fluid	0.001 Pa s
ν	Poisson's ratio of matrix frame	0.2

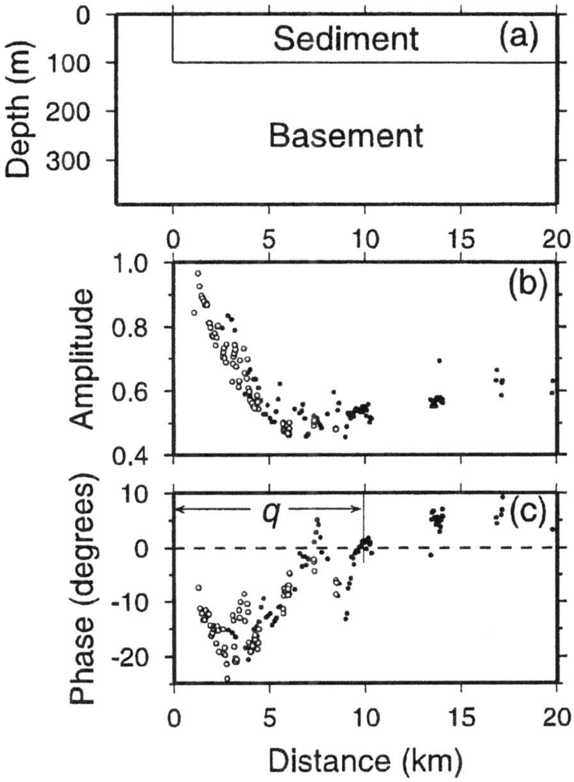

Figure 11. (a) Simplified geometry for interpreting the observed tidally induced formation fluid pressure variations. The formation responds instantaneously to loading directly from above, but the diffusive signal propagates horizontally into the permeable layer from the basement outcrop. (b) Observed amplitude and phase of formation fluid pressure variations with tides, recorded by ODP holes 1024C (open circles) and 1025C (solid circles). All periods are presented as their 12-hr period equivalent at various distances as explained in the text.

penetration lengths of these waves in the low-permeability sediment section. Although the diffusive pressure wave comes only horizontally from the western end, the formation is loaded elastically directly from above by ocean tides. The elastic component can be equivalently described as being due to the same loading function applied horizontally with no vertical deformation, because the elastic pore fluid pressure change depends solely on the volumetric deformation of the matrix frame, not on the direction of compression. Therefore, our system is equivalent to a slice of a half-space turned sideways, with the western end being its "upper" boundary. The solution shown in Figure 10 applies, with depth z changed to horizontal distance x.

The amplitude and phase of the formation pressure observed in ODP holes 1024C and 1025C relative to the observed seafloor loading are shown in Figures 11b and 11c, respectively, as functions of distance from outcrop. The spectrum represents a broad band of frequencies including oceanographic and barometric, as well as tidal, pressure variations. An example of the actual pressure time series from 1024C has been given in Figure 3a. In diffusion-wave propagation, distance x scales with the square root of the tidal period, that is, $x / \sqrt{F}$ appears as a single variable in the mathematical expressions (e.g., Wang et al., 1998). Therefore, observing a pressure wave of period F_1 at distance x_1 is equivalent to observing a wave of period F_2 at distance $x_2 = x_1\sqrt{F_2/F_1}$. Pressure variations of many different tidal and non-tidal periods recorded in the two boreholes have been converted using this relation in order to represent a wave of 12-hour period observed at many different distances. The observed phases and amplitudes shown in Figures 11b and 11c can thus be compared with the theoretical curves in Figures 10a and 10b, respectively. The amplitudes at large x values give a tidal loading efficiency of 0.6, which can be used to infer the elastic properties of the formation (Davis et al., 2000). The zero cross-over of the phase function gives the half-wave length q of the diffusive pressure wave. From Figure 11b, we find $q \approx 10$ km. Boreholes 1026B and 1027C (Figure 1) are located at distances much greater than this q value, and the diffusive component of the dominant diurnal and semi-diurnal tidal signal is not expected to reach them. Observations from these holes verified this point; pressure amplitudes in these holes do not vary much with period, and phases are close to zero (Davis et al., 2000).

Substituting $q = 10$ km and $F = 12$ hrs into (8), we obtain a hydraulic diffusivity of 740 m^2/s. The value of μ and parameter values used to derive ζ' are given in Table 2. Compared to the permeability, these parameters are much better known. Applying these η, μ and ζ' values to (6), Davis et al. (2000) obtained a permeability of 1.7×10^{-10} m^2, in good agreement with the estimates based on thermal observations and modeling described in Section 2.

Having been equipped with the knowledge of poroelasticity, we can revisit the question of how well the formation pressure was monitored using a pressure sensor in a sealed borehole. In the introduction, we implied that a good measurement requires the formation permeability to be high. This is because of the compressibility of the fluid in the borehole (i.e., borehole storage). In order for a pressure variation in the permeable formation to be transmitted into the borehole, fluid must flow between the formation and the borehole. An increase (decrease) in formation pressure drives the fluid into (out of) the borehole. If the formation permeability is low, the borehole pressure will lag behind the formation pressure. Boreholes 1026B and 1027C, the two boreholes that are not affected by diffusive tidal signals from the ridge area, provide a good test for the borehole storage effect. The fact that tidal pressures in these holes show practically zero phase lag relative to seafloor loading indicates that there is good hydraulic communication between the borehole and the permeable formation and that the borehole storage effect is negligible. In a different setting where formation permeability may not be as high, the effect of borehole storage should be taken into account. However, a new design of the CORK system that is currently being tested allows formation pressures to be measured in small isolated sections of the borehole and thus effectively minimizes the borehole-storage effect.

4. PRESSURE VARIATIONS AND FLUID FLOW INDUCED BY TECTONIC-STRAIN

The tidal variations discussed above are the dominant signals in the formation pressure records. By removing these signals from the recorded time series, we are able to examine the more subtle but distinct pressure variations due to changes in tectonic strain. For example, if a fault suddenly slips, it causes a stress change around it in the poroelastic medium, and the stress change induces pore fluid pressure change. This is similar to tidal loading, but the loading is not from the seafloor and is not periodic.

At the time of or immediately after a tectonic strain event, the formation is "undrained", and the fluid responds only elastically. In a 1-D process, the diffusive component does not become important for an observation spatial scale L until the time lapsed is comparable to the diffusion time constant τ as defined in (7). In 3-D, the diffusion takes an even longer time. Given a total pressure change σ_t in the poroelastic medium, the elastic (undrained) fluid pressure change p is

$$p = \gamma\sigma_t \tag{9}$$

and the product of p and β_f gives the undrained volumetric strain of the fluid. To know the spatial distribution of σ_t given a tectonic loading event, we need to know the volumetric strain change θ of the matrix frame which can be provided by an elastic deformation model. A volume change of the frame is caused by the pressure change that effectively works on the frame only: the effective pressure (σ_t - αp), where α, often assumed to be 1, is defined in Table 1. According to Hooke's law, the relation between the effective pressure (and hence the pore fluid pressure) and frame volumetric strain is

$$\theta = \beta(\sigma_t - \alpha p) = \beta(\frac{1}{\gamma} - \alpha)p \tag{10}$$

Figure 3b shows pressure variations caused by a tectonic-strain event in three of the ODP holes along our transect (Figure 1). Hole 1026B had a leakage problem in its CORK, and in addition its pressure record at the time of the tectonic event was affected by another ongoing experiment using this hole. As a result, the pressure record from 1026B was too noisy to be used to resolve the tectonic signal. As explained in detail by Davis et al. (2001), the observed strain pulse was due to a largely aseismic spreading event at the Juan de Fuca Ridge. As constrained by the accompanying seismicity and the co-event pressure change in another borehole to the north along the ridge, our borehole transect is located right along the line of symmetry of the ridge segment where the spreading event took place. The spreading event caused a compression of the crust along the transect and induced a fluid pressure increase in the permeable formation. The elastic response, represented by a sudden pressure rise at the beginning of the event, occurred in all three sites simultaneously, with magnitude decreasing with increasing distance from the ridge. The subsequent continuing rise and eventual decay of the fluid pressure are consequences of Darcy flow that depend on the formation permeability.

The strain field (of the matrix frame) associated with the spreading event can be derived using a model of "dilatational dislocation" in a uniform elastic half-space (Okada, 1992). We assumed a Poisson's ratio of 0.25 for the half-space; the 3-D dislocation solution does not require other elastic moduli to be defined. The fluid pressure induced by the elastic strain can be obtained using (10) and parameters in Table 2. The best model that gives a good fit to the initial (assumed undrained) pressure rise observed at all three locations suggests a widening of 12 cm along a 40-km long ridge segment to 3 km depth (Davis et al., 2001). The geometrical details of the deformation source area are unimportant to the pressure field a few kilometers away. The plan view of the initial pressure field from this model is shown in Figure 12a. The dislocation model is 3-D, but the volumetric strain and thus the instantaneous fluid pressure is calculated at the surface of the half-space, consistent with the fact that the depth extent of our formation pressure

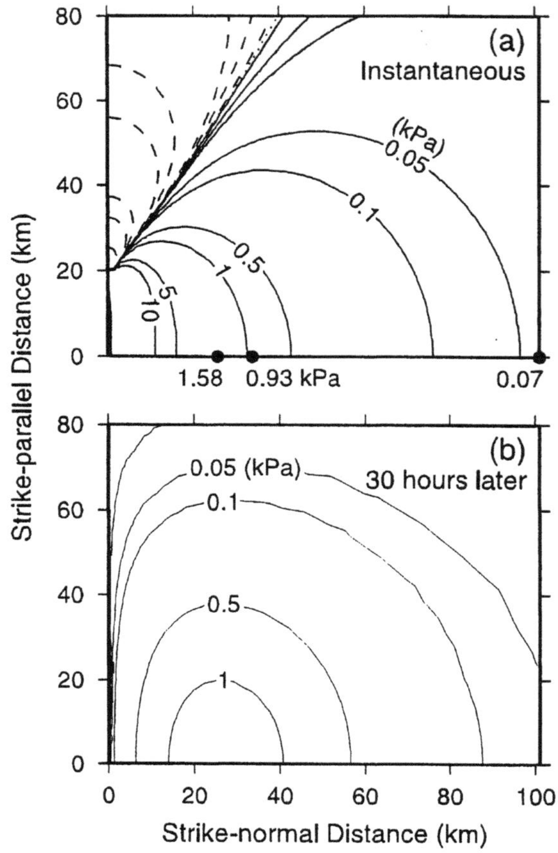

Figure 12. (a) Elastic fluid pressure increase due to a spreading event at the Juan de Fuca Ridge, modeled as a 0.12 m dilatation of a ridge segment between -20 km and 20 km. The pressure has been converted from the volumetric strain calculated using a 3-D dislocation model assuming $p = 9 \times 10^9 \theta$ (consistent with parameter values in Table 2). Borehole locations (solid circles) and the observed initial pressure increase (numbers below the solid circles) are also shown for comparison. (b) Pressure field 30 hours after the initial pressure increase as predicted by a 2-D Darcy flow model. For both (a) and (b), the symmetric solution in the southern half of the model is not shown.

observations is much smaller than the dimension of the source area of the tectonic loading.

The continuing pressure change after the initial response is mostly the result of diffusive Darcy flow. As the flow takes place, the fluid pressure change should modify the deformation of the solid matrix frame, which in turn will affect the fluid pressure, although the effect of this deformation-flow coupling is very small. We ignore this coupling and employ a simple 2-D finite element Darcy flow solution using as an initial condition the coseismic pressure distribution predicted by the dislocation model (Figure 12a). The flow model assumes that water flows only horizontally in the highly permeable volcanic formation of the crust and that the permeability is uniform. Given the simple layered structure of the oceanic crust and the very low permeability of the sediment cover, these assumptions are reasonable for a first-order model.

Two mechanisms generate pressure gradients that drive horizontal flow in the formation. The first is the spatial variation of the initial pressure field. The elastic contraction is greater, and hence the fluid pressure is higher, near the source region of the spreading event along our borehole transect. The other mechanism is drainage from the unsedimented area around the source region, the same place where ocean tides generate the diffusive pressure wave ("e" in Figure 8c). Any anomalous formation fluid pressure such as that caused by the tectonic strain will cause water to flow into or out of the highly fractured formation through this exposed region. Darcy flow driven by the initial pressure gradient causes the pressure "front" to move away from the source region, that is, eastward along the transect. The pressure at a given borehole location thus continues to rise after the initial (elastic) increase. However, rapid drainage through the basement exposure to the west gives rise to a westward pressure gradient. As this "negative" pressure front propagates eastward, the borehole site that initially experiences the "post-elastic" pressure increase will eventually experience a pressure decrease. The farther away from the source region, the later will both the pressure maximum and subsequent decay occur. To simulate the drainage effect, we prescribe a zero pressure increase along the ridge axis boundary. Other model boundaries, set very far from the source region, are not affected by the spreading event, so they are maintained at zero pressure as well.

We have calculated the pressure field evolution using different hydraulic diffusivities. The time scale for diffusion at each of the sites provide constraints on the large-scale hydraulic diffusivity of the formation. To reproduce the primary characteristics of the pressure variations for all the three borehole sites, that is, the rate of the post-elastic increase, the time of the peak value, and the rate of the subsequent decay, requires a diffusivity of about 2500 m^2/s. The model-predicted pressure histories at the borehole sites are shown in Figure13, and the plan view of the pressure field at 30 hours since the beginning of the spreading event is shown in Figure 12b. With (6) and parameter values given in Table 2, the formation permeability is estimated to be between 10^{-9} and 10^{-10} m^2. The range represents mainly the uncertainties that arise from the simplifying assumptions of the model. This permeability estimate is consistent with those inferred from the thermal observations and tidal pressure records. It is appropriate over a lateral scale of many tens of kilometers.

5. SUMMARY

It is well known that permeability is a property that depends on scale. Young oceanic crust is particularly heterogeneous, and it is reasonable to expect the scale dependence to be extreme. Permeabilities determined by pumping tests in deep-ocean boreholes are sensitive to a scale of tens to hundreds of meters and are found to be several orders of magnitude greater than those measured in core samples. We have reviewed three novel strategies that provide permeability determinations that are sensitive to an even greater scale (km to tens of km), and they indeed yield values that are much higher than the borehole determinations. The three strategies include:

(1) Determining the degree to which the thermal regime of sediment-buried oceanic crust is non-conductive: With local basement topography, sediment thickness variation, and uppermost basement thermal structure constrained by seismic reflection, seafloor heat flow, and drilling data, the degree of thermal homogenization of the upper oceanic crust created by buoyancy-driven fluid circulation can be determined, and hence the permeability can be estimated. In one well-studied

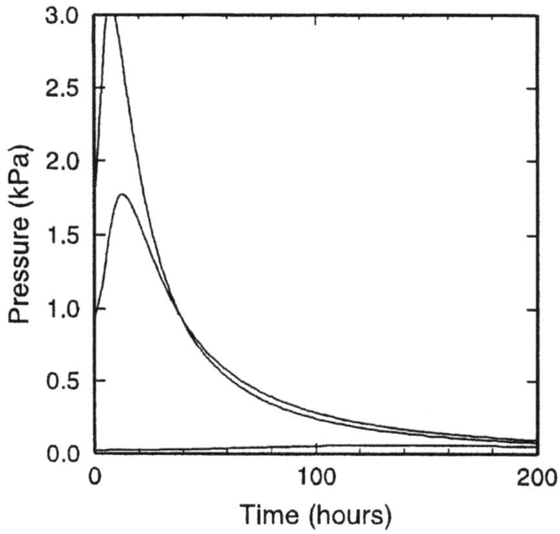

Figure 13. Model-predicted pressure variations with time at the three borehole sites. The abrupt rise at time zero is the elastic response. The characteristics of the curves should be compared with observations shown in Figure 3b. From Davis et al. (2001).

example on the eastern Juan de Fuca Ridge flank, the uppermost crust in a continuously sedimented area of high basement relief is observed to be uniform to within 2 K over a distance of kilometers, despite a 2.5:1 variation in sediment thickness.

(2) Determining the frequency dependence and scale of the diffusion of seafloor pressure variations: This strategy has been applied on the eastern Juan de Fuca Ridge flank at a location where sediments onlap the elevated basement topography near the ridge axis. Seafloor and associated formation pressure variations are observed over a broad range of periods from a few hours to several weeks. Diurnal variations are seen to propagate in the crust beneath the sediment cover laterally as a diffusion wave with a penetration length of 10 km.

(3) Determining the dissipation rate following instantaneous changes in pressure generated internally by tectonic strain: Strain associated with seismic and aseismic tectonic events cause step-wise changes in interstitial fluid pressure. Fluid flow follows the pressure gradients that are generated, and the pressure history is controlled primarily by permeability. This phenomenon has been observed at three sites on the Juan de Fuca Ridge flank, and the permeability constrained by the observations provides an average over a scale of several to several tens of km.

Permeability estimated using all three of these strategies are in the range of 10^{-10} m^2 to 10^{-9} m^2, one to three orders of magnitude higher than the range derived from borehole observations in the same region (Fisher et al., 1997; Becker and Fisher, 2000). It is clear that young oceanic crust is capable of hosting huge fluid flux and associated heat and geochemical fluxes even where pressure driving forces are small. The permeability values obtained at the Juan de Fuca Ridge flank may be representative of other young oceanic crusts with a similar age, although it is expected that the permeability will decrease with increasing crustal age as a result of the precipitation of crack-filling minerals. Applying the strategies summarized in this paper in regions of older crust should allow this process to be investigated. Unfortunately, none of the strategies provides a constraint on the depth to which high permeability extends. For the purpose of estimating the degree to which crustal rocks can be altered and ultimately carry water into subduction zones, this is an important matter. Few boreholes penetrate more than the uppermost few tens of meters into young ocean crust. Those that do suggest that permeability decreases rapidly with depth, although the scale dependence of permeability may be even more extreme at deeper levels, where water-bearing fractures are probably spatially rare but may dominate the hydrologic regime. New strategies to approach the questions of depth and distribution of permeability are required.

Acknowledgments. We thank the numerous colleagues who contributed to the studies reviewed in this article, particularly, K. Becker and R. MacDonald for contribution to the development of the CORK system, G. van der Kamp for sharing knowledge of poroelasticity, R. E. Thompson for tidal pressure time series analysis, and J. He for developing numerical techniques for modeling hydrothermal convection. Comments from the two anonymous referees improved the paper. Geological Survey of Canada contribution 2001095.

REFERENCES

Becker, K., Fisher, A. T., 2000. Permeability of upper oceanic basement on the eastern flank of the Juan de Fuca Ridge determined with drill-string packer experiments. J. Geophys. Res. 105, 897-912.

Clauser, C., 1992. Permeability of crystalline rocks. Eos Trans. AGU 73, 233, 237-238.

Davis, E. E., Chapman, D. S., Forster, C. B., Villinger, H., 1989. Heat-flow variations correlated with buried basement topography on the Juan de Fuca Ridge flank. Nature 342, 533-537.

Davis, E. E., Becker, K., Pettigrew, T., Carson, B., Macdonald, R., 1992. CORK: A hydrologic seal and downhole observatory for deep ocean boreholes. Proc. Ocean Drill. Program Init. Rep. 139, 43-53.

Davis, E. E., Wang, K., He, J., Chapman, D. S., Villinger, H., Rosenberger, A., 1997. An unequivocal case for high Nusselt number hydrothermal convection in sediment-buried igneous oceanic crust. Earth Planet. Sci. Lett. 146, 137-150.

Davis, E. E., Wang, Becker, K., K., Thomson, R. E., 2000. Formation-scale hydraulic and mechanical properties of oceanic crust inferred from pore pressure response to periodic seafloor loading. J. Geophys. Res. 105, 13,423-13,435.

Davis, E. E., Wang, K., Thompson, R. E., Becker, K., Cassidy, J. F., 2001. An episode of seafloor spreading and associated plate deformation inferred from crustal fluid pressure transients. J. Geophys. Res. 106, 21,953-21,963.

Fisher, A. T., 1998. Permeability within basaltic oceanic crust. Rev. Geophys. 36, 143-182.

Fisher, A. T., Becker, K., Davis, E. E., 1997. The permeability of young oceanic crust east of Juan de Fuca Ridge determined using borehole thermal measurements. Geophys. Res. Lett. 24, 1311-1314.

Giambalvo, E. R., Fisher, A. T., Martin, J. T., Darty, L., Lowell, R. T., 2000. Origin of elevated sediment permeability in a hydrothermal seepage zone, eastern flank of the Juan de Fuca Ridge, and implications for transport of fluid and heat. J. Geophys. Res. 105, 913-928.

Houtz, R., Ewing, J., 1976. Upper crustal structure as a function of plate age. J. Geophys. Res. 81, 2490-2498.

Keenan, J. H., Keyes, F. G., Hill, P. G., Moore, J. G., 1978. Steam Tables. John Wiley, New York, 162 pp.

Kirby, S., Engdahl, E. R., Denlinger, R., 1996. Intermediate-depth intraslab earthquakes and arc volcanism as physical expressions of crustal and physical expressions of crustal and uppermost mantle metamorphism in subducting slabs. In: Bebout, G. E., Scholl, D. W., Kirby, S. H., Platt, J.P.(Eds.), Subduction: Top to Bottom, AGU Monograph 96, 195-214.

Lister, C. R. B., 1990. An explanation for the multivalued heat transport found experimentally for convection in a porous medium. J. Fluid Mech. 214, 287-320.

Okada, Y., 1992. Internal deformation due to shear and tensile faults in a half space. Bull. Seismol. Soc. Am. 82, 1018-1040.

Peacock, S. M., 2001. Are the lower planes of double seismic zones caused by serpentine dehydration in subducting oceanic mantle? Geology, 29, 299-302.

Peacock, S. M., Wang, K., 1999. Seismic consequences of warm versus cool subduction metamorphism: Examples from Southwest and Northeast Japan. Science 286, 937-939.

Van der Kamp, G., Gale, J. E., 1983. Theory of Earth tide and barometric effects in porous formations with compressible brains. Water Resour. Res. 19, 538-544.

Rohr, K., 1994. Increase of seismic velocities in upper oceanic crust and hydrothermal circulation in the Juan de Fuca plate. Geophys. Res. Lett. 21, 2163-2166.

Snelgrove, S. H., Forster, C. B., 1996. Impact of seafloor sediment permeability and thickness on off-axis hydrothermal circulation: Juan de Fuca Ridge eastern flank. J. Geophys. Res. 101, 2915-2925.

Wang, K., Davis, E. E., 1996. Theory for the propagation of tidally induced pore pressure variations in layered subseafloor formations. J. Geophys. Res. 101, 11,483-11,495.

Wang, K., He, J., Davis, E.E., 1997. Influence of basement topography on hydrothermal circulation in sediment-buried igneous oceanic crust. Earth Plan. Sci. Lett. 146, 151-164.

Wang, K., Davis, E. E., van der Kamp, G., 1998. Theory for the effects of free gas in subsea formations on tidal pore pressure variations and seafloor displacements. J. Geophys. Res. 103, 12,339-12,353.

Wang, K., van der Kamp, G., Davis, E. E., 1999. Limits of tidal energy dissipation by fluid flow in subsea formations. Geophys. J. Int. 139, 763-768.

Wilkens, R. H., Fryer, G. J., Karsten, J., 1991. Evolution of porosity and seismic structure of upper oceanic crust: Importance of aspect ratios. J. Geophys. Res. 96, 17,981-17,995.

INDEX OF AUTHORS

SUBJECT INDEX

A

B

C

D

E

euphotic zone 156, 159
excess radon 155

F

G

H

N

O

P

R

S

www.ingramcontent.com/pod-product-compliance
Ingram Content Group UK Ltd.
Pitfield, Milton Keynes, MK11 3LW, UK
UKHW050945290926
15470UKWH00005B/1642